Design Risk Management
Contribution to Health and Safety

Design Risk Management Contribution to Health and Safety

Stuart D. Summerhayes
BSc, MSc, CEng, MICE, RFaPS

A John Wiley & Sons, Ltd., Publication

This edition first published 2010

Blackwell Publishing was acquired by John Wiley & Sons in February 2007. Blackwell's publishing programme has been merged with Wiley's global Scientific, Technical, and Medical business to form Wiley-Blackwell.

Registered office
John Wiley & Sons Ltd, The Atrium, Southern Gate, Chichester, West Sussex, PO19 8SQ, United Kingdom

Editorial offices
9600 Garsington Road, Oxford, OX4 2DQ, United Kingdom
2121 State Avenue, Ames, Iowa 50014-8300, USA

For details of our global editorial offices, for customer services and for information about how to apply for permission to reuse the copyright material in this book please see our website at www.wiley.com/wiley-blackwell.

Library of Congress Cataloging-in-Publication Data

Summerhayes, Stuart.
Design risk management: contribution to health and safety / Stuart D. Summerhayes.
p. cm.
Includes bibliographical references and index.
ISBN 978-1-4051-3275-6 (pbk. : alk. paper) 1. Construction industry–Safety measures. 2. Safety factor in engineering. 3. Architecture–Human factors. I. Title.
TH443.S86 2010
690'.22–dc22
2009030831

A catalogue record for this book is available from the British Library.

Set in 11 on 14 pt Plantin by Toppan Best-set Premedia Limited
Printed and bound in Malaysia by KHL Printing Co Sdn Bhd

1 2010

Contents

1	**Introduction**		**1**
	Table 1.1	Design failures	5
	1.1	Major design failures in British history	7
	1.2	Additional Reports (The Bragg Report and HSE Research Report 218) into design failure	12
	Table 1.2	Principal recommendations of the Bragg Committee	15
	Table 1.3	Contributory factors to historical failures	19
2	**Project risk management and design risk management**		**23**
	2.1	Key players in project management	26
	2.2	Stages of the contract and their achievement	28
	Table 2.1	CDM duty holder actions	31
3	**Construction-related health and safety legislation**		**35**
	3.1	Approved code of practice and guidance	38
	3.2	Health and Safety at Work, etc. Act 1974	39
	3.3	The Management of Health and Safety at Work Regulations 1999	40
	3.4	The Workplace (Health, Safety and Welfare) Regulations 1992	40
	3.5	The Manual Handling Operations Regulations 1992	41
	3.6	The Confined Spaces Regulations 1997	42
	3.7	The Work at Height Regulations 2005	42
	3.8	The Control of Vibration at Work Regulations 2005	43
	3.9	The Control of Noise at Work Regulations 2005	43
	3.10	The Control of Substances Hazardous to Health Regulations (amended 2005)	44
4	**The CDM process**		**45**
	4.1	Timing	47
	4.2	Pre-construction information	47
	Figure 4.1	Holistic diagram of the construction process	49
	4.3	Construction phase plan	51
	4.4	Health and safety file	51
	Figure 4.2	Systems approach	53
	Table 4.1	Applicable regulations for duty holder compliance	55
	Table 4.2	Construction (Design and Management) Regulations 2007	57
5	**Role of the designer**		**61**
	5.1	Who are designers?	63
	Figure 5.1	The designer's duties	65
	Table 5.1a	Designer duties (all projects)	69
	Table 5.1b	Designer duties (additional duties on notifiable projects)	71

6 The design risk management process 75
6.1 Additional interfaces 81
6.2 Design change 83

7 Documentation 85
Table 7.1 Risk assessment methods 91
Table 7.2 Examples of potential hazards for designers to consider 95
7.1 Red, amber and green lists 98
Table 7.3 Design risk assessment 101
Figure 7.1 Example of a design risk assessment proforma 103
Figure 7.2 Annotated notes (health and safety) on drawing 107
Figure 7.3 Hazard management register and design risk assessment 109
7.2 Project (health and safety) risk register 111
Table 7.4 Project risk register (health and safety) 113
7.3 Design philosophy statements 115

8 Information flow 119
Table 8.1 Communication links 123
8.1 Pre-construction information 126
Figure 8.1 Information flow 127
Figure 8.2 Design interface with other duty holders 131
8.2 Construction phase plan 133
8.3 Health and safety file 134
Table 8.2 Design information for the health and safety file 137
Figure 8.3 Planning/programming integration 141

Appendix One: Roadmap 143
Appendix Two: References and bibliography 145
Appendix Three: Web page directory 149
Appendix Four: Workplace (Health, Safety and Welfare) Regulations 1992 151
Appendix Five: Design checklist 155
Appendix Six: Riba Outline Plan of Work 2007 (November 2008 revision) 159

Index 163

A colour plate falls between pages 74 and 75

Section 1
INTRODUCTION

Much of the aesthetic appeal of design is achieved through the endeavours of design teams who seek to convert the abstract into the functional, compatible with the laws of science, arts and nature. Whilst our landscape has been shaped and marked by the many iconic 'state-of-the-art' flagship projects, other challenging design solutions pass into the shadow of the infrastructure support that many of us take for granted.

The post-industrial era, particularly, has provided opportunities for the designer to explore the limitations of understanding and provide solutions that stand as evidence of beauty, scale, functionality and purpose. That we can still enjoy the benefits of their appreciation and vision centuries after their completion says much about the position occupied by the designer, whether as architect, engineer or design consultant. Society's debt to designers is well documented and establishments continue to laud and honour their many achievements.

The names of Wren, Telford, Stephenson, Rennie, Tarmac and Brunel still illuminate past horizons, whilst Arup, Foster, Rogers and others light up the path from the present into the future. Whilst the continuum from Wren to Rogers covers the period of modern history, much has changed in the use of design in the modern environment. Materials and processes have altered, fashioned by technological advances. Concepts and philosophies have evolved to encompass client involvement and the legal framework has been developed to impart further responsibility and ownership on individuals and the design team generally.

Regardless of such change, good design has always embraced health and safety issues, but it is the visibility and transparency of this outcome that is now different. The modern challenge to designers is no more limiting than in the past and many would argue that there are now greater opportunities for designers to use their creativity in addressing the health and safety implications associated with their designs.

Design insight and technical advances are unfortunately built on the foundations of design failure, and it is to the credit of our predecessors that the painful lessons learnt have given rise to procedural, technical and managerial improvements, and the delivery of more effective and robust project management regimes. Aligned to this situation is the perspective now demanded that designers comprehensively appreciate the health and safety discharge of duties, which represents the modus operandi of this book.

Whilst an appreciation of design success provides the necessary perspective that leads to progress and also facilitates an understanding of design evolution, an analysis and appraisal of design *failure*, as tabulated in Table 1.1, offers a further vehicle to the necessity of change, allowing us to learn in equal measure and move forward, aware of the need to avoid the mistakes of the past. Such movement must also acknowledge the sanctity of human life, which is enshrined in the Health and Safety at Work, etc. Act 1974 (hereafter HSAW 1974).

The extension of and challenge to technical boundaries has never been incompatible with the demands of a 'safe and suitable' working environment. Prestigious buildings can offer constructability, operability, maintainability and replaceability all within the acceptable framework of health and safety.

Table 1.1
DESIGN FAILURES

Year	Description	Structure	Fatalities	Comments
1879	River Tay, Scotland	Bridge failure	75 (no survivors)	Court of enquiry made significant criticisms of the design engineer
1925	Llyn Eigiau, Dolgarrog, Wales	Dam	16	Led Parliament to pass the Reservoirs (Safety Provisions) Act in 1930
1966	Aberfan, Wales	Colliery spoil tip	144	Tribunal led by Lord Justice Davies
1968	Ronan Point, Canning Town, London	High rise flats – gas explosion	4 (17 injured)	Introduced concept of disproportionate collapse and brought about fundamental changes to the design philosophy of building structures in the UK
1970	Cleddau, Milford Haven, Wales	Box girder bridge collapse	4	Merrison Committee set up and prepared Interim Design and Workmanship Rules (IDWR)
1972	River Loddon	Bridge collapse	3 (10 injured)	*Interim report of the Advisory Committee on Falsework* (HMSO, 1974) (The Bragg Report) and subsequently BS 5975 (1982) *The Code of Practice for Falsework*
1973	Summerland, Isle of Man	Leisure complex fire	50 (50 injured)	Enquiry commission identified many human errors and failures and some very ill-defined and poor communication
1973	Camden School, London	Sports hall roof collapse	None	Conclusions included lack of bearing, lack of reinforcement continuity, rebar corrosion and conversion of high alumina concrete Current codes of practice reviewed
1974	Flixborough, Yorkshire	Chemical plant explosion	28 (36 injured)	Shortcomings identified in the official enquiry led to significant tightening of the UK government's regulations covering hazardous industrial processes
1984	Abbeystead	Pumping station explosion	16 (28 injured)	HSE report made numerous recommendations in respect of design and construction as well as operational management
1985	Valley Parade, Bradford, Yorkshire	Football stadium fire	56 (265 injured)	Popplewell Inquiry (1986) resulted in new legislation governing safety at sports grounds around the UK
1988	Piper Alpha, North Sea	Oil platform fire and explosion	167	Public enquiry chaired by Lord Cullen (Report 1990) Offshore Installations (Safety Case) Regulations 1989
1989	Hillsborough, Sheffield	Football stadium	96	*Final report into the Hillsborough Stadium Disaster* (HMSO, 1990), (The Taylor Report)
1994	Port of Ramsgate	Passenger walkway collapse	6 (7 injured)	Inquiry identified numerous areas where lessons needed to be learnt
1994	Heathrow, London	Tunnel collapse	None	Recovery took 2 years and cost around £150 million
1998	Docklands Light Railway, Lewisham	Tunnel collapse	None	BS 6164 (*Code of practice for safety in tunnelling in the construction industry*) amended
1999	Avonmouth, Bristol	Maintenance platform failure	4	Kvaerner (Clevelend Bridge) Ltd and Costain Ltd fined £500,000 each and costs of £525,000

1.1 Major design failures in British history

Table 1.1 is not an exclusive list but does catalogue a number of high profile historical design failures that have been the subject of forensic analysis and the basis of procedural and statutory change.

A further insight is offered below into a selection of these cases and one other (Nicholls Highway Project) in order to appreciate mechanisms of failure and to provide a rationale towards an understanding of the procedural controls now implicit in achieving compliance with current construction related legislation.

Tay Bridge disaster[1]

The collapse of the Tay Bridge, with the accompanying loss of life, cast a shadow over Victorian engineers.

- Designed for North British Railway by the engineer Sir Thomas Bouch (1822–1880).
- Eighty-five spans, 13 of which were navigation spans. Eleven of these were 245 feet long and two were 227 feet long; the remainder of bridge spanned between 67 feet and 164 feet.
- Contract went to the lowest bidder – Charles de Bergue – and was completed by Messrs Hopkins, Gilks & Co. of Middlesborough due to the personal illness of Charles de Bergue.
- Tender price was £217,099 18 s 6 d.
- Disaster occurred on Sunday 28 December 1879.
- Evening train from Edinburgh to Dundee consisting of one engine and six carriages crossed onto bridge at 07:14 pm in the teeth of a strong westerly gale of 60–0 mph (Beaufort scale of between 10 and 11).
- Driver had no warning as train ploughed off bridge; engine found with throttle fully open.
- No survivors; 75 dead.

Rothery, the wreck commissioner, saw fit to publish his own report in which he wrote:

> *The conclusion then, to which we have come, is that this bridge was badly designed, badly constructed and badly maintained ... For these defects both in the design, the construction and the maintenance Sir Thomas Bouch is, in our opinion mainly to blame. For the faults in design he is entirely responsible. For those of construction he is principally to blame in not having exercised that supervision over the work which would have enabled him to detect and apply a remedy to them. And for the fault of maintenance he is also principally, if not entirely to blame in having neglected to maintain such an inspection over the structure as its character imperatively demanded.*

[1] *The Report of the Court of Enquiry*, W. Yoland and W.H. Barlow (1880). *An Addendum to the Enquiry Report*, H.C. Rothery (1880).

Summerland disaster[2]

On the evening of 2 August 1973, a fire started outside a leisure complex on the Isle of Man close to one of the walls and spread to the interior. The building quickly became engulfed in fire and all floors at and above entrance level were completely destroyed. The majority of the 3000 people within escaped, but 50 people perished, with a similar number treated in hospital. At the time, in terms of loss of life, this was the worst peace-time disaster in the British Isles since 1929 (Glen Cinema Disaster, Paisley – 71 deaths). The casualty numbers were attributable to the rapid development of the fire and the delayed evacuation of the building.

Some of the report conclusions were as follows:

- *No efficient design management was applied ... It is a design team's responsibility to consider carefully the functions of a building, particularly from the point of view of its efficient usage, comfort, maintenance safety. Elsewhere the Commission has been critical, not so much of the part choice of certain materials, but of the way they were used, with little understanding of their limitations.*
- *No one ever stood back and looked at the project as a whole.*
- *The motive (of Trust House Forte Leisure Ltd) was the earliest opening date, but the procedures verged on the irresponsible.*

Some of the recommendations included the following:

- *In the designing of a building a named person should be in charge from the outset and take and be known to be taking the major design decisions.*
- *Architects and clients should together carefully consider the requirements and performance of a building-in-use at the stage when conceptual designs are proposed.*
- *Architectural training should include a much extended study of fire protection and precautions.*
- *A set of detailed and up-to-date plans of the premises, showing the essential structure and services, should be available in all occupied buildings.*

The Abbeystead explosion[3]

Designed as part of the Lancashire Conjunctive Use Scheme (Lune–Wyre Transfer Scheme) to meet the expected water supply requirements of the area, the Lune–Wyre transfer link comprised the Lune Intake and Screenhouse, the Lune Pumping Station, the Quernmore Pipeline, the Wyresdale Tunnel and the Abbeystead Outfall Station.

- Security, environmental considerations and protection of valves against freezing dictated that the proposed valve-house building at Abbeystead should be largely underground.
- Design enquiries suggested limited geological information based on Ordnance Survey geological maps made in 1870s.

[2] *Report of the Summerland Fire Commission*, Government Office, Isle of Man (1974).
[3] *The Abbeystead Explosion*, Health and Safety Executive, HMSO (1985).

- Obtaining further information via drilled bore holes was considered, but only a few were actually drilled.
- The decision to limit borehole information was supported by an independent specialist.
- Routine conditions prevailed.
- Traces of flammable natural gas were detected during tunnel driving, but contractors and consulting engineers regarded the tunnel to be gas-free by normal tunneling standards.

Important characteristics in relation to the Abbeystead explosion were:

- All the contents of the tunnel, both liquid and gaseous, discharged into a room with limited natural ventilation.
- Water passed through a concrete lined tunnel, i.e. a tunnel not designed to be watertight.
- Ground water from the strata surrounding the tunnel leaked in rather than tunnel water leaking out.

Client:	North West Water Authority
Designer:	Binnie and Partners
Contractor:	Edmund Nuttall Limited
Commencement:	End of 1975
Completion:	Spring 1979
Contractual responsibilities:	Ended 15 December 1980

On the evening of Wednesday 23 May 1984, between 1830 and 1900 hours, a party of 44 people, including 8 employees, was assembled at the Abbeystead Valve House (the visit was to address residents' concerns that water pumped into the Wyre at Abbeystead had aggravated local flooding).

- Prior to the visit no water had been pumped for 17 days; it was intended to pump during the visit as a demonstration.
- A telephone call was made at about 1912 hours for pumping to start at the supply end. After 10 minutes, after no water flowed, a further telephone call was made and the order given to start up second pump.
- An explosion occurred at around 1930 hours.
- Eight people died at the scene, eventually rising to 16 people; no one escaped without injury.
- Substantial damage caused to the valve-house.

The explosion was caused by ignition of a mixture of methane and air, which had accumulated in the wet room of Abbeystead Valve House. No source of ignition for the explosion has been positively identified.

Numerous recommendations in respect of design and construction and operational management were highlighted in the Health and Safety Executive (HSE) Report.

HSE prosecuted. After an appeal Binnie and Partners were found to be 100% responsible.

Port of Ramsgate ferry disaster[4]

Shortly before 0100 on the evening of Wednesday 14 September 1994, part of the passenger walkway at No 3 Berth at the Port of Ramsgate collapsed. One end of the walkway fell 10 metres, embedding itself in the deck of the pontoon that had provided the floating seaward support for the structure. Six members of the public were killed and seven received multiple injuries.

Client:	Port of Ramsgate
Designer:	Fartygsentrepenader AB
Contractor:	Fartygstionstructioner AB
Approval organisation:	Lloyds Register of Shipping

In early 1994 the single-deck Berth 3 linkspan at Port Ramsgate was substantially modified to provide an upper deck with a new upper vehicle bridge, and a separate high-level walkway was installed to lead from a new shore building to the passenger deck of a ferry. On 28 April 1994, the completion certificate for the Berth 3 upper-deck project was signed. The passenger walkway was brought into use on 12 May 1994.

Review of the design revealed that it did not provide the support and articulation necessary to match the overall design concept. The walkway was designed in such a way that it was likely to be torsionally stiff. As such, the design did not allow for the roll of the pontoon and the design calculations of the loadings on the cantilevered support stub axles were inadequate. It appeared that the designers had failed to visualise how the static and dynamic loadings would be carried and therefore failed to consider the effects of fatigue on the support stub axles. No fatigue calculations were made. Additionally, no provision was made for continuing maintenance of the upgraded structure, lubrication facilities were not installed, suitable access for maintenance was not incorporated in the design and no manual or other written instructions were provided.

The report concludes that the collapse was caused by a series of errors in the design, some of which were gross. Underlying the mechanical causes of the collapse were the failures of major parties engaged in the project to carry out their respective functions adequately.

In particular, there was:

- a failure of any of the parties to carry out a risk assessment for the project allowed safety-critical design failures to be made
- the failure to have a project plan that provided for the effective monitoring of the project allowed defects in design and fabrication to remain undetected.

Even when defects became apparent to certain individuals, the lack of adequate systems of liaison and communication prevented effective action being taken to remedy them and, more importantly, prevented any fundamental consideration of a series of defects and problems which might have led to the questioning of the underlying technical causes of these defects.

[4] *Walkway Collapse at Port Ramsgate. A Report on the Investigation into the Walkway Collapse at Port Ramsgate on 14 September 1994*. Health and Safety Executive, HMSO (2000).

Among the lessons learned were the need for:

- promotion of effective project management
- competent design and fabrication
- adequate maintenance information
- proactive risk assessments
- effective communication.

All were convicted of serious offences under HSAW 1974 and record fines and costs (£2.4 million) were imposed.

Heathrow Express tunnelling project[5]

The tunnels collapsed in the early hours of Friday 21 October 1994 and continued to fail over a number of days. Although there was no loss of life or injury the failure brought chaos to the heart of Heathrow Airport.

Client:	British Airports Authority
Main contractor:	Balfour Beatty Civil Engineering
Tunnelling consultant:	Geoconsult

Described by the HSE as 'one of the worst civil engineering disasters in the last quarter of a century'.

As well as criticising 'poor construction' the report also underlined the following:

- '*breaking the link between design of permanent and temporary works created difficulties in taking an integrated design approach to risk reduction*'
- a catalogue of design and management errors, poor workmanship and quality control were at the root of the catastrophic tunnel collapse
- errors were made leading to poor design and planning, a lack of quality control during construction, a lack of engineering control and most importantly a lack of safety management
- '*risk assessment should be a fundamental step in the procedures adopted by all parties: it is inappropriate wholly to leave the control risk to contractors*'
- '*those involved in projects with the potential for major accidents should ensure they have in place the culture, commitment, competence and health and safety management systems to secure the effective control of risk and the safe conclusion of the work*'
- '*collapse could have been prevented but for a cultural mindset which focused attention on the apparent economies and the need for production rather than the particular risks*'.

Outcome:

- Balfour Beatty was fined £1.2 million for two offences under HSAW 1974.
- Geoconsult, the tunnelling consultant, was fined a further £500,000 plus £100,000 costs.

[5] *The Collapse of NATM Tunnels at Heathrow Airport*, HSE Books (2000).

- The total fine of £1.7 million was a record at the time for offences under health and safety legislation.

Nicholls Highway tunnel collapse, Singapore[6]

This catastrophic collapse occurred on 20 April 2004 on a section of cut and cover tunnel built under contract C824 for Singapore Metro's new Circle Line. Excavation of the 15–20 m-wide trench had reached 30 m below ground level when retaining walls gave way, caving in over a 110 m length. As a result four workers died.

Client:	Singapore Land Transport Authority
Joint venture partners:	Nishimatsu and Lum Chang
Sub-contractors:	Numerous
Procurement	Design and build

Findings include:

- lack of continuity between design and construction
- failure to apply the same safety factors to temporary works as to permanent works
- lax safety culture
- engineers failed to address properly the risks of low probability and high magnitude accidents because they had not seen them occur before.

Report recommendations included:

- a balancing of production measures against safety measures
- the provision of a temporary works designer responsible for checking design and the installation of temporary works
- attention to the performance of non-standard designs.

It was observed that contractual complexity with poor definition of responsibilities and inadequate lines of communication combined with lack of interaction between designers and constructors were key factors in both this collapse and the Heathrow Express collapse.

1.2 Additional Reports (The Bragg Report and HSE Research Report 218) into design failure

Many of the criticisms identified by the above cases are further endorsed in both the Bragg Report[7] (see below) on falsework collapses and the HSE Research Report 218[8], which looked at causation effects of site accidents.

[6] *New Civil Engineer*, 23 September 2004.

[7] *Final Report of the Advisory Committee on Falsework*, Health and Safety Executive, HMSO (1976).

[8] *Peer Review of Analysis of Specialist Group Reports on Causes of Construction Accidents*, Research Report 218, Health and Safety Executive, HMSO (2004).

The Bragg Report

The aim of the Bragg Committee was to find out why accidents associated with falsework/formwork collapses occurred and to recommend how they might be avoided, with particular reference to the collapses at Loddon Viaduct (23 October 1972: three men killed and ten others injured), Birling Road overbridge (23 March 1971: one man killed, five men seriously injured and twelve others slightly injured) and similar accidents in Europe, the Middle East, Canada, Australia and America.

Studies showed that there were multiple causes for the failures, but that each failure composed of two elements: the technical cause that led to collapse and the procedural errors that allowed the faults to occur and to go undetected and uncorrected.

The principal recommendations are outlined in Table 1.2.

Table 1.2

PRINCIPAL RECOMMENDATIONS OF THE BRAGG COMMITTEE

	Description	Procedure
5.	In his calculations the designer should allow for possible variations in positioning and alignment, which are inevitable even with good workmanship. The drawings should state the tolerance within which the falsework must be constructed.	Constructability
6.	All falsework must be *designed*, even if on a small job the design is only a sketch. The designer, especially if he is not on site, must have a proper written brief, which must include all the factors that might have to be allowed for.	Systematic and disciplined approach
10.	Suppliers of proprietary materials should be required to specify the conditions of test, the failure loads and the mode of failure of each item of equipment in addition to any recommendations about safe working loads.	Material limitations and supply chain information
11.	Tests should be carried out on new materials to check the validity of claims made for them and on used materials to check the deterioration which occurs in service.	Quality assurance
12.	The designer should assume that previously used material will be incorporated in falsework and must use appropriate stresses. If there are critical areas where he has assumed the use of new material these must be clearly indicated on drawings.	Communication
15.	The falsework design and, if he requests them, the calculations that were made must be submitted to the designer of the permanent works for comment. If the person responsible for the permanent works is an architect without engineering qualifications he must submit them to his consulting engineer unless the building method is traditional in all respects.	Design interface between temporary and permanent
16.	The philosophy of preparing and checking the design, of not modifying it without assessing the resulting effects and of having any doubtful points checked must apply in all cases, major and minor.	Co-ordination
17.	On all sites the contractor or construction organisation must appoint a properly qualified temporary works co-ordinator whose duties are to ensure that all procedures have been followed, that all checks and inspections have been carried out and that any modifications or changes have been properly authorised. Falsework may not be loaded or struck without the written permission of the temporary works co-ordinator.	Co-ordination, competence and ownership
18.	Communication between designers and others on and off site must be improved. Drawings must be clear and loading diagrams must be provided.	Team integration and communication

Research Report 218

Research Report 218 identified that:

- Regulations need to be read in conjunction with the relevant Approved Code of Practice.
- Designers have a vital contribution to health and safety matters on all projects. Obviously theirs is not the only contribution but it is an influential and critical contribution as an integrated part of the health and safety management team.
- The cultural shift places emphasis on designs that are safer and healthier to build; operate; maintain and demolish.
- Many designers remain intransigent and fail to embrace the challenge.
- The challenge is for a radical change from within the design fraternity.
- 'The Report concludes that almost all accidents in construction could have been prevented by designer intervention and that at least 1 in 6 of all accidents are at least partially the responsibility of the lead designer in that opportunities to prevent accidents were not taken.'
- Designing from the health and safety perspective of construction workers continues to be one of the challenges of delivering good design.
- It is at the conceptual design stage that many decisions are taken that irrevocably shape the construction process. In the early stages of design effective health and safety management can influence the entire process and contribute to added value through commercial viability.
- Health and safety is an issue that has to be managed through the design process and on-site. It has to become a management, not a medical, issue if the industry is to prevent ill-health. The construction industry's safety culture is a collective commitment to safety.
- 'The final numbers are not just persuasive but absolutely convincing. Designers can do more.'
- The risk-tolerant culture of the construction industry, including that among clients and designers, must be changed.
- Cost, not safety, cannot remain the culture. Construction is price and not quality driven despite the initiatives since the Latham report.
- Paper-chase bureaucracy is not the fault of regulations but of those who abdicate managerial duties and fail to make decisions about what is relevant and what is not.
- Simply completing a documentary record and reviewing it is inadequate and unhelpful.
- Something like 60% of accidents have their roots upstream of what happens on the construction site.
- The designer, like other construction professionals, has moral, professional, financial and statutory obligations to be fulfilled in the discharge of design duties.
- Further accident prevention could have occurred by design intervention (43%) or by having a temporary works designer (1 in 6).

Thankfully there has been change, but there are still lessons to be learnt and whilst the criticisms contained within the above incidents/reports can over-shadow us all, they should simply serve as a reminder of the seriousness of the business of construction and the need for constant vigilance in ensuring that procedures and processes achieve their intended objective.

Failure is rarely uni-causal and therefore all duty holders have a contribution to make. This is no more apparent than in the role and function discharged by the design team, who are high up in the supply chain and invariably function as the professional adviser to the client (and others).

Every effort must be made to adopt a pro-active integrated team response and avoid the spectre of repeat situations, but it should not be assumed that all the lessons from past occurrences have been embedded into the designer's psyche. Procedures and controls must constantly challenge any suggestion of complacency.

The role of the designer implicitly confers a consummate need in matters of health and safety management to *contribute* arising from sufficient *consideration* of associated hazards, coupled with due *communication* within a *co-operative* and integrated team framework. The attainment of these objectives can be thwarted singularly or collectively by the spectre of the fragmented team, complacency, professional arrogance or the 'radar screen of awareness' being switched off.

As noted in paragraphs 109 and 110 of the Approved Code of Practice (ACoP)[9] to the Construction Design and Management Regulations 2007(CDM Regulations 2007):

> *'Designers are in a unique position to reduce the risks that arise during construction work, and have a key role to play in CDM Regulations 2007. Designs develop from initial concepts through to a detailed specification, often involving different teams and people at various stages. At each stage, designers from all disciplines can make a significant contribution by identifying and eliminating hazards, and reducing likely risks from hazards where elimination is not possible.*
>
> *'Designers' earliest decisions fundamentally affect the health and safety of construction work. These decisions influence later design choices, and considerable work may be required if it is necessary to unravel earlier decisions. It is therefore vital to address health and safety from the start.'*

Design teams are therefore key players as well as essential contributors and communicators in matters of health and safety management, and each team member must acknowledge that ineffectiveness in either is a precursor to failure in both project success and the discharge of statutory duties.

The following table, Table 1.3, summarises further some of the contributory mechanisms that have led to historical failure. They should all forewarn design teams of vigilance towards the avoidance of complacency.

[9] *Managing Health and Safety in Construction,* Approved Code of Practice (L144), HSC, HSE Books (2007).

Table 1.3
CONTRIBUTORY FACTORS TO HISTORICAL FAILURES

Category	Detail	Reference
Procedural	No training manual	Avonmouth Bridge
	Permit type system identified in risk assessment was not implemented	Avonmouth Bridge
	Lack of monitoring in respect of procedural compatibility	Avonmouth Bridge
	Lack of safety in design and construction	Avonmouth Bridge
	Allocation of responsibilities unclear	Cleddau Bridge
	Safety procedures inadequate	Hillsborough Stadium
	Outdated safety certificates	Hillsborough Stadium
	Inadequate 'permit to work' management systems	Piper Alpha
	Ineffective safety management systems	Piper Alpha
	Communication failure	Piper Alpha
	Safety audits-ineffective	Piper Alpha
	Lack of emergency planning	Piper Alpha
	No efficient design management applied	Summerland
	Lack of holistic overview	Summerland
	Ill-defined and poor communications	Summerland
	Failure to carry out a risk assessment	Port of Ramsgate
	Roles poorly understood	Heathrow Tunnel
	Lack of engineering control	Heathrow Tunnel
Technical	Lack of control (physical stops)	Avonmouth Bridge
	Inadequacy of design of a pier support diaphragm. Bridge design and construction code of practice was inadequate for such application.	Cleddau Bridge
	Inadequate fire protection	Piper Alpha
	Design simulation failure	Piper Alpha
	Design failure (mechanical)-lack of appreciation	Flixborough
	Discrepancies in sub-structure information	Tay Bridge
	Design inadequacies	Tay Bridge
	Inferior workmanship	Tay Bridge
	Use of materials – lack of understanding	Summerland
	Inadequate site investigation	Aberfan
	HSE investigation failed to reveal any calculations had been carried out for the overburden pressure along the tunnel	Docklands Light Railway
	Design faults	Ronan Point
	Design faults	Cleddau Bridge
Organisational	Lack of training	Avonmouth Bridge; Piper Alpha
	Health and safety reports not acted upon	Avonmouth Bridge
	Inadequate levels of supervision	Tay Bridge
	Poor communication and failure	Tay Bridge
	Lack of adequate systems of liaison and communication	Port of Ramsgate
	Lack of communication	Aberfan
	No attempt to evacuate the 3000 people present until visible evidence of the flames prompted a panic-stricken rush for the exits, where many people were crushed and trampled.	Summerland
	Failure of site organisation between the parties	Cleddau Bridge
	Poor workmanship and inspection procedures	Ronan Point
	'All the hallmarks of an organisational accident'	Heathrow Tunnel
Managerial systems	No effective management response to previous incidents	Avonmouth Bridge
	Poor design and planning	Heathrow Tunnel
	Minimal qualifications	Piper Alpha
	Poor practices and ineffective audits	Piper Alpha
	Lack of provision of effective monitoring	Port of Ramsgate
	Lack of safety management	Heathrow Express
	A catalogue of design and management errors	Heathrow Express

It is vital that the issues associated with the list of failures remind us all of the link between the legacy of the past and the challenges that lie ahead. For both the design and project team, the outcomes of forensic analysis must ensure that our systems of control provide a route whereby duties are effectively discharged, that health and safety hazards are satisfactorily managed and that adherence to the process is demonstrable without being excessively burdensome. This is the thrust of the Construction (Design and Management) Regulations 2007.

Successful attainment is dependent on the calibre of design managers, the competence of the design team, the efficacy of process control and the commitment to continual improvement in all design matters affecting project outcomes.

Section 2

PROJECT RISK MANAGEMENT AND DESIGN RISK MANAGEMENT

It is no surprise that the review exercise that delivered the Construction (Design and Management) Regulations 2007 sought closer alignment with the process of project management. Project management itself has historically developed and provided many of the procedures and controls that are now deemed essential for successful project delivery.

Project management is a dynamic process that utilises the appropriate resources of the organisation in a controlled and structured manner in order to achieve clearly defined objectives identified as strategic needs. It is always conducted within a defined set of constraints. There are numerous definitions of project management, including that from the *Code of Practice for Project Management for Construction and Development* published by the Chartered Institute of Building, which provides the following definition:

'Project Management may be defined as the overall planning, control and co-ordination of a project from inception to completion aimed at meeting a Client's requirements in order that the project will be completed on time within authorised cost and to the required quality standards.'

It is no coincidence that project management terms are liberally sprinkled throughout the CDM Regulations 2007, for both project management and health and safety risk management are implicitly combined. Both rely on the integrated team concept for success and each represents a 'journey of constant improvement'. Indeed, health and safety risk management has always been an integral sub-set of project management irrespective of any legislative connection.

To that end the project management terms which remind all duty holders of the need for a proactive, ongoing approach resonate throughout the regulations, as shown below.

Action required	Duty holder	Regulation
'... maintained and reviewed ...'	Client	9(2)
'... shall ensure ... are complied with throughout the construction phase'	Contractors	13(7)
'... revised as often as may be appropriate ...'	Client	17((3)(b)
'... take appropriate action ... where it is not possible to comply ...'	Contractor	19(2)(b)
'... planned, managed and monitored ...'	Principal contractor	23(1)(a)
'... made and implemented ...'	CDM co-ordinator	20(1)(b)
'... review and update ...'	CDM co-ordinator	20(2)(e)
'... review, revise and refine ...'	Principal contractor	23(91)(b)

Fundamental to this is the role of the designer, who is not only close to the start of the supply change but also occupies an influential position in both health and safety management contribution and communication.

Project management in its simplest form is about the management of change and combines the synergy of the team members into a dynamic but temporary and well-focused arrangement, based on a *proactive* approach, in order to achieve the *project aims*. This utilises the essentials of *risk management* with adequate controls enabling prompt action to be taken to ensure deliverables and targets are appropriately met.

Such definitions and objectives are almost inter-changeable with those of design risk management.

The success of project management delivery is often spoken of in terms of 'on time; on cost and on quality', but this is too simplistic and does not account for the subjectiveness of those who are measuring. Perhaps a better definition is the delivery of projects 'on time, on cost, on quality, with everyone's reputation enhanced'. This cannot be achieved without an approach that is duly compliant to the requirements of health and safety legislation.

Project failure is experienced by most construction professionals at some stage of their career and always presents another opportunity for improvement and development. Such failure is usually measured in terms of project over-runs or budgetary excesses. However, failure in health and safety management is inevitably accompanied by a reduction in the quality of someone's life and all too often in the death or serious injury of that individual.

The causation effects of project management failure include inadequate project definition, infrastructure confusion, lack of ownership, scope creep, incompetence, complacency, reactive management, unrealistic resource appreciation, over-optimism, monitoring failure, team fragmentation and ineffective communication and so on.

These factors also stalk health and safety risk management, and the concept of the integrated team is a fundamental pre-requisite for all projects in the delivery of effective project management and a pre-requisite for the delivery of health and safety compliance under the CDM Regulations 2007 and all other construction-related workplace legislation. Team fragmentation undermines the very basis of the team approach within the holistic concept and contributes significantly to communication failure and lack of process control. This is unacceptable on professional, moral and legislative grounds.

Project management as a collective term embraces other management facets, including: risk management, time management, and health and safety management. All are inextricably linked, with failure in one leading inevitably to failure in others. Additionally, the consequence of health and safety management failure evidenced by the fatality and occupational ill-health statistics remains a specific indictment of project team ability and the culture of the construction industry at large.

2.1 Key players in project management

The public sector model identifies the following key players in project management.

Project steering group

This is the project board or cabinet and is made up of the senior managers who are responsible for the strategic input and for the effective transition across functional boundaries that is essential for success of the matrix management model. In essence they represent the client for whom the project is carried out.

Project sponsor

The project sponsor is sometimes referred to as the project director and is accountable to the project steering group for the overall performance of the project. The sponsor drives the project in the right direction for the benefit of the organisation. It should be a singular role with full accountability.

Project manager

This is the key role and is filled by the individual or organisation responsible to the sponsor for the day-to-day management of the project and who is charged with completing the work on time, to an agreed budgeted cost and quality. The project manager, therefore, has the duty of providing a cost-effective and independent service correlating, integrating and managing different disciplines and expertise to satisfy the objectives and provisions of the project brief from inception to completion.

The role involves the supply of technical expertise to assess, procure, monitor and control the resources needed to complete the project and it is essential that everyone involved in project management accepts that the project manager is expected to cut across normal organisational boundaries to get the job done.

Project team

These are the individuals (consultants, contractors, specialists and others) who carry out all the tasks planned in the project schedule. The team comprises core team members and extended team members, the latter being part of the team for a limited period of time. The project team share responsibility for delivery of the project execution plan. Individually each member of the team is responsible to the project manager for the tasks they have been assigned. The majority of CDM duty holders reside in this category, e.g. designer(s), CDM co-ordinators, contractor(s) and the principal contractor.

Resource managers

These are departmental managers with direct responsibility for supplying team members for the project.

Stakeholders

This includes all those with an interest in the project (positive and negative), whose views must be taken into account during the development of the project.

Customer

This is the end user or 'purchaser' of the project outcomes or results, which may include the public. It should be noted that not all stakeholders share similar aspirations for the project. *What is crucial is that they all share similar and acceptable aspirations for the outcomes of health and safety management.*

However, note that in respect of terminology, under the engineering and construction contract the project manager is one of two 'employer's representatives'.[1]

Herein, the project manager's role is to manage the contract on behalf of the employer with the intention of achieving the *employer's* objectives (usually expressed in terms of a budget, a programme and a brief setting out the requirements for the end-product). His authority under the contract is expressed in terms of the actions that the contract prescribes to him and includes authority to change the works information, to instruct the contractor to do various things and to generally exercise his managerial and engineering judgement.

The project manager's actions with respect to quality include:

- giving instructions to change the works information
- giving instructions to resolve ambiguities or inconsistencies between the documents that are part of the contract
- acceptance of the contractor's design for any parts of the works and for items of equipment
- acceptance of replacement of key people
- acceptance of sub-contractors, the proposed subcontract conditions (in certain circumstances) and the proposed subcontract date (in certain circumstances)
- acceptance of method statements and the rest of the programme
- giving instructions to stop or not to start any work
- deciding the date of and certifying completion of the project
- potentially, acceptance of the contractor's quality policy statement and quality plan if additional clauses are added to take this into account
- potentially giving instructions to the contractor to correct a failure to comply with the quality plan if additional clauses are added to take this into account
- assessing amounts payable by the contractor to the employer in respect of costs incurred by the employer and others resulting from a test or inspection having to be repeated after a defect is found
- arranging for the employer to give access to the contractor to parts of the works already taken over by the employer, if this is needed to correct defects
- proposing changes to the works information so that defects do not have to be corrected
- arranging to have defects corrected by people other than the contractor where the latter has not corrected them within the defect correction period
- requesting proof of the contractor's title to documents, equipment, plant and materials prior to inclusion of the value of these in assessments of the amount due.

Hence, it is to be appreciated that under such forms of contract the project manager has great authority and responsibility, and could well have an impact on design and design development.

2.2 Stages of the contract and their achievement

Each project can be broken down into definable key stages.

[1] Mitchell B and Trebes B, *NEC Managing Reality Book 3: Managing the Contract*, Thomas Telford Books (2005).

1. Getting started (conception)
   - Establish business case
   - Nominate the project sponsor
   - Review options
   - Obtain professional advice
   - Develop strategy
   - Select project manager
2. Defining the project
   - Establish infrastructure
   - Develop the strategic brief
   - Develop the project execution plan
3. Assembling the team
   - Select the project team
   - Decide contracts
   - Control of the project team
     - bi-weekly project review meeting
     - weekly cost and procurement meetings
     - prompt, clear client decisions
     - accessibility
     - single-point responsibilities
     - social occasions
4. Designing and constructing
   - Develop the project brief
   - Develop concept design
   - Develop detailed design
   - Incorporate control systems
   - Start construction
   - Manage and resolve problems
   - Review progress and quality
   - Account for testing and commissioning
5. Completion and evaluation
   - Ensure project is capable of 'going operational'
   - Complete the project
   - Undertake post-project audit – post-project evaluation
     - Performance against cost, quality and timescale targets, etc.
     - Client and user satisfaction
     - Overview and recommendations
   - Evaluate feedback.

Compare the list above with the RIBA plan of works (Appendix 6).

To achieve the steps outlined above, project management must be proactive and integrated, with appropriate levels of planning, response and organisational control exercised throughout the project period. Such controls also exist within the CDM model, where they are to be discharged by the appropriate duty holder (see Table 2.1).

Table 2.1

CDM DUTY HOLDER ACTIONS

Regulation	Duty holder	Action
4	Client	Competence
5		Co-operation
6		Co-ordination
7		General principles of prevention
9		Management arrangements
12		Appointment of designers outside G.B
14		Appointments
16		Control over start of construction
4	Designer	Competence
5		Co-operation
6		Co-ordination
7		General principles of prevention
11(1)		Awareness of duties
11(2)(3) and (4)		Risk management
12		Appointment of designers outside G.B
18(1)		Commencement of significant design stage
4	CDM co-ordinator	Competence
5		Co-operation
6		Co-ordination
7		General principles of prevention
20(1)(b)		Co-ordination of health and safety
20(1)(c)		Liaise with principal contractor
20(2)(c)		Ensure designers comply with their duties
20(2)(d)		Ensure co-operation between designers and principal contractor
21		Ensure notification goes off
4	Contractor	Competence
5		Co-operation
6		Co-ordination
7		General principles of prevention
13(1)		Awareness of duties
13(2)		Plan, manage and monitor
13(6)		Site security
13(7)		Welfare
19(1)		Start of construction work
19(2)		Risk assessments
19(3)(a) and (b)		Review of construction phase plan
19(3)(c)		Notify principal contractor
4	Principal contractor	Competence
5		Co-operation
6		Co-ordination
7		General principles of prevention
22(1)(a)		Plan, manage and monitor
22(1)(b)		Liaise with CDM co-ordinator
22(1)(c)		Ensure welfare facilities are sufficient
22(1)(d) and (e)		Rules and directions
22(1)f)		Minimum time for planning/preparation
22(1)(g) and (i)		Provision of information
22(1)(l)		Site security
22(2)(a), (b), (c)		Induction, information and training
23		Construction phase plan
24		Consultation

The challenge for project management in the classical sense is that it is concerned not with the continuing operation of an organisation, but with the effective action by a disparate group of people who have different objectives but who have been harnessed to a common goal, albeit temporarily. In this sense there is a limited time for which the synergy of such a team can be maintained. However, as with CDM management, it is dependent on the integrated team for successful completion taking due cognisance that successful projects are delivered by people with the right attitude. The synergy of such a team cannot be allowed to dissipate on any project.

Hence there are similarities between project management and design risk management, with both aligned to the concept of the integrated team sharing common goals and seeking to continually improve. The design risk management process in matters of health and safety must also embrace the general principles of prevention to the satisfaction of the term, 'as far as is reasonably practicable'.

Consider the parallels shown below.

Project management	**Design risk management**
Key components:	**CDM compliance:**
Clear objectives	Section 3 (HASW Act 1974)
Organisational and individual relationships (Latham and Egan principles)	Integrated and holistic team
Communication	Regulations 11(6) and 18(2)
Competence	Regulation 4
Co-ordination and effective interfacing	Regulation 5
Co-operation and collaborative working	Regulation 6
Controls	Regulations 7, 11(1) and 18(1)
Control methodology	Design risk management
Cultural alignment	Paragraph 3 (Approved Code of Practice) and Section 1 of HSAW 1974

Project planning and organisation together with risk and contingency management are integrated subsets of both project management and design risk management, and are key elements in delivering success on all projects. The sooner the pertinent issue is identified, the greater the time available to address the issue and mitigate against the outcome. This can only be achieved by proactive responses and appropriate ownership by the team member (duty holder).

An effective project management/design management organisational structure sets out unambiguously and in detail how the parties to the project are to perform their functions in relation to each other in contributing to overall success. It also identifies protocols and procedures for monitoring and control, as well as outlining relevant administrative details. Such infrastructure arrangements are critical for success and are facilitated by the concept of the *lead designer* as referred to in paragraph 20 of the ACoP.

An effective organisational structure must also have the ability to respond to the dynamic challenge presented by resource optimisation, within and between design teams and throughout projects.

The organisational model must be updated as circumstances dictate during the project lifetime and should allow project objectives to be communicated and agreed by all concerned for the promotion of effective teamwork. This is the role of both project manager and/or lead designer, and is based on proactivity, monitoring, revision and communication. Such duties are set out in the CDM Regulations 2007 and other construction related legislation, and clarified in the associated Approved Codes of Practice.

> *'Effective briefing is essential throughout the project. However, perhaps the most important element is the time spent at the outset.'*
>
> *Briefing the Team*, Thomas Telford, Working Group 1, CIB (1997).

This is true of project management generally, with many notable failures attributable to lack of communication and a failure to adequately develop the client brief. An ill-defined project is rarely capable of successful delivery, with a premature start devoid of effective procedures and control mechanisms simply contributing to the breakdown of team effectiveness, as parties seek to compromise. An identical scenario exists for design risk management.

Whilst one of the boundary conditions for all projects is budgetary restraint, it is helpful to appreciate the typical owning costs of a building[2], which are stated as being in the ratio of:

- 1 for construction costs
- 5 for maintenance and building operating costs
- 200 for business operating costs.

For medium-sized to large projects, this cost profile promotes the need for a lifecycle cost approach to evaluate the true cost of constructing, using and operating such an asset.

It also serves to remind us of the delusion in considering the cost of a project at any other intermediate point other than obsolescence. For the designer, it poses the question of enhanced specification and more construction cost to achieve a maintenance/ replacement strategy compatible with compliance to the *general principles of protection* (Regulation 7).

Since the design team occupies a critical supply chain position as professional advisor to the client, they should be able to exert great influence over that one duty holder who influences many of the budgetary decisions that impact on health and safety outcomes emanating from the design process. Such outcomes impact on:

- specifications
- installations
- maintenance methodologies
- service provisions.

It is not difficult, accounting for the Royal Academy of Engineering's cost ratio, to present the argument for health and safety management compliance in business case terms rather than solely in health and safety compliance terms.

[2] *The Long Term Costs of Owning and Using Buildings*, Report of the Royal Academy of Engineering (1998).

Section 3
CONSTRUCTION-RELATED HEALTH AND SAFETY LEGISLATION

The design team not only occupies a crucial supply chain position but usually operates as professional advisors to the client. This enables them to exert great influence over the client, the one duty holder who carries more legislative responsibility than any other under the CDM Regulations 2007 and it is to the lead designer that the client will inevitably turn for guidance in the early days of most projects. Historically and contractually the client/design interface is robust and contact with a designer is often the first point of contact for the lay client in seeking to progress with project development for whatever purpose.

All design teams must ensure that the client is aware of his duties on all construction projects (Regulation 11(1)), but this is a duty inevitably undertaken by the lead designer, with the other designers seeking re-assurance from him that this has been done. Additional duties are then derived for the notifiable project at the demarcation point between 'initial' and 'significant detailed'[1] design (Regulation 18(1)). It is worth noting that both these statutory duties are to be undertaken in advance of the appointment of the CDM co-ordinator.

Designers therefore must have the relevant process awareness and detailed knowledge in respect of not only the CDM Regulations 2007 but also all other legislation pertinent to that construction sector operation. Additionally, whilst Regulations 11 and 18 of the CDM Regulations 2007 outline the modus operandi of design process management that is relevant to health and safety within construction, it is apparent that a working understanding is also required of other duty holder roles as well as corresponding legislation in specific areas of design involvement and interest. This understanding supports the interface management required between designer/client, designer/contractor(s), designer/CDM co-ordinator and designer/principal contractor.

Such a working knowledge must be embraced by the design team and the following have some relevance on the majority of projects, but are not exclusive:

- Health and Safety at Work, etc. Act 1974
- Management of Health and Safety at Work Regulations 1999.
- Workplace (Health, Safety and Welfare) Regulations 1992
- Manual Handling Operations Regulations 1992.
- Confined Spaces Regulations 1997
- Work at Height Regulations 2005
- Control of Vibration at Work Regulations 2005
- Noise at Work Regulations 2005
- Control of Substances Hazardous to Health Regulations 2005.

The HSAW etc Act 1974, which was the catalyst for the modernisation of workplace legislation in the aftermath of the Robens Report, is often referred to as both primary legislation and as an enabling act, whilst the remainder of the above list constitutes a list of secondary legislation.

The Act's enabling facilities included the formation of the:

- Health and Safety Commission
- Health and Safety Executive
- Employment Medical Advisory Service

as well as the publication of:

- approved codes of practice
- guidance notes

[1]Paragraph 66, ACoP.

and the establishment of the political structure to allow subsequent legislation to become part of the statutory framework without going through the full parliamentary process, thus offering an accelerated route.

The hierarchy of legal documents is laid out as follows:

- statutory legislation, e.g. Construction (Design and Management) Regulations 2007
- approved codes of practice, e.g. L144, *Managing Health And Safety In Construction*, HSE Books (2007)
- guidance documents, e.g. HSG 65, *Successful Health and Safety Management*, HSE Books (1997).

To appreciate the detail of any Regulation it is essential to read that Regulation in conjunction with the corresponding approved code of practice, which provides interpretation and clarification facilitated by pragmatic examples. However, it is the literal interpretation that would form the basis of judicial judgement in the event of any ambiguity. The legal position as well as the purpose of any approved code of practice for workplace legislation must be appreciated by the construction professional as outlined below.

3.1 Approved code of practice and guidance

The following extracts from L144, the approved code of practice for *Managing Health and Safety in Construction. Construction (Design and Management) Regulations 2007*, outline the purpose, scope and main duties created by the regulations. This document has been widely acclaimed within the industry and follows a user-friendly format. It is an essential reference for every duty holder, who should note the significance and status of such a document.

> *'However, the Code has a special legal status. If you are prosecuted for breach of health and safety law, and it is proved that you did not follow the relevant provisions of the Code, you will need to show that you have complied with the law in some other way or a Court will find you at fault.'*

It follows therefore that each duty holder needs to have the awareness and detail of what is required in compliance with the discharge of his duties, as well as a working knowledge of how other duty holders with whom he or she is interfacing within the project situation should be performing.

The Approved Code of Practice (ACoP) further notes that:

> *'The key aim of CDM 2007 is to integrate health and safety into the management of the project and to encourage everyone involved to work together to:*
>
> - *improve the planning and management of projects from the very start;*
> - *identify hazards early on, so that they can be eliminated or reduced at the design or planning stage and the remaining risks can be properly managed;*
> - *target effort where it can do the most good in terms of health and safety; and*
> - *discourage unnecessary bureaucracy.'*

It is important that all those who can contribute to the health and safety of a construction project understand what they and others need to do under the Regulations and discharge their responsibilities accordingly. This will require a radical change in culture for many of the duty holders as well as training and education in the practical steps and procedures to be taken for compliance. Design risk management systems could well be scrutinised as a result of a non-compliant approach associated with another duty holder who is contractually involved in the project.

Paragraph 121 of the ACoP notes that:

'These (design) duties apply whenever designs are prepared which may be used in construction work in Great Britain. This includes concept designs, competitions, bids for grants, modifications of existing designs and relevant work carried out as part of feasibility studies. It does not matter whether or not planning permission or funds have been secured; the project is notifiable or high- risk; or the client is a domestic client.'

This can now be extended to include Northern Ireland, where almost identical legislation was introduced in July 2007.

3.2 Health and Safety at Work, etc. Act 1974

The Health and Safety at Work, etc. Act was the outcome of the earlier Robens report (1972) and marked the modernisation of workplace legislation. It was an enabling act and moved the approach away from prescriptive legislation to one of a 'deemed to satisfy' or a 'management by objectives' approach. It is *criminal law* and furthermore invokes a reverse burden of proof, in that any duty holder accused of a breach of health and safety law is presumed guilty until they can prove their innocence.
Together with the formation of the HSC, HSE, EMAS and the publication of approved codes of practice, another of the Act's enabling features is the creation of secondary legislation without the need to go through the full parliamentary process.

Specifically, Section 3 places general duties on all employers and the self-employed towards persons other than their employees, namely:

'3-(1) It shall be the duty of every employer to conduct his undertaking in such a way as to ensure, so far as is reasonably practicable, that persons not in his employment who may be affected thereby are not thereby exposed to risks to their health or safety.'

Section 6 places obligations on those who design, manufacture, import or supply any article or goods for use at work, to ensure that such equipment is safe to install, use and maintain without risks to health and safety and requires the provision of relevant information to achieve these objectives.

Section 37 deals with offences caused by the body corporate and as such carries a serious message for any director.

'37-(1) Where an offence under any of the relevant statutory provisions committed by a body corporate is proved to have been committed with the consent or connivance of, or to have been attributable to any neglect on the part of, any director, manager, secretary or other similar officer of the body corporate or a person who was purporting to act in any such

capacity, he as well as the body corporate shall be guilty of that offence and shall be liable to be proceeded against and punished accordingly.'

(2) Where the affairs of a body corporate are managed by its members, the preceding subsection shall apply in relation to the acts or defaults of a member in connection with his functions of management as if he were a director of the body corporate.'

It reinforces the concept of *duty of care*, which can be used by the HSE to achieve a successful conviction when the specifics of other legislation cannot be relied upon and lays the foundation for the right of all employees to come to work, under *safe and suitable* conditions and therefore go home safely at the end of each working day.

3.3 The Management of Health and Safety at Work Regulations 1999

The Management of Health and Safety at Work Regulations 1999 is a revision of the earlier 1992 Regulations, which arose as the principal method of implementing the EC Framework Directive (83/391/EEC). The duties here overlap with other Regulations because of their wide-ranging nature. Central to its effect on all those involved in the workplace was the introduction via Regulation 3 of the legal requirement to undertake 'suitable and sufficient' assessment of workplace risks.

Regulation 3 Risk Assessment

'Every employer shall make a suitable and sufficient assessment of:

(a) The risks to the health and safety of his employees to which they are exposed whilst they are at work; and
(b) The risks to the health and safety of persons not in his employment arising out of or in connection with the conduct by him of his undertaking.'

Regulation 4 outlines the hierarchical principles of prevention qualified by the term 'so far as is reasonably practicable', which should be familiar to designers through their decision-making response to health and safety contribution. These principles are also contained in Schedule 1 of these Regulations, but perhaps are now more familiar to designers through Appendix 7 of the ACoP (L144) and also the pragmatism associated with Regulations 11(4) and 11(6) of the CDM Regulations 2007.

3.4 The Workplace (Health, Safety and Welfare) Regulations 1992

The Workplace (Health, Safety and Welfare) Regulations 1992 identify the basic requirements of fixed workplaces such as offices, shops, factories and schools but do not include construction sites. They stem from the general duty on employers under the HSAW Act to ensure the health, safety and welfare at work of all their employees.

As noted in paragraph 2...

'These Regulations apply to a very wide range of workplaces, not only factories, shops and offices but, for example, schools, hospitals, hotels and places of entertainment. The term workplace also includes the common parts of shared buildings, private roads and paths on industrial estates and business parks, and temporary work sites (but not construction sites).'

For the designer the importance lies in his duties under Regulation 11(5) of the CDM Regulations 2007.

'In designing any structure for use as a workplace the designer shall take account of the provisions of the Workplace (Health, Safety and Welfare) Regulations 1992 which relate to the design of, and materials used in, the structure.'

The designer of retaining walls, bridges, underpasses and pipeline distribution systems would not have to comply with these Regulations provided there was no fixed workplace associated with their usage on completion.

3.5 The Manual Handling Operations Regulations 1992

The Manual Handling Operations Regulations 1992 tackle the largest category (musco-skeletal disorders) of occupational ill-health in the UK and require all those with a responsibility to eliminate manual handling 'as far as is reasonably practicable'. This single issue has been the focus of numerous HSE initiatives, who have always informed that there is no such thing as a safe manual handling operation. Relevant parties, including designers, must look to mechanisation and automation to eliminate the need for manual handling and the focus required gets to the heart of many constructability issues.

As noted in the introduction to the HSE guidance document *Backs for the Future* (HSG 149):[2]

'This guidance will, therefore, be of interest to:

- *clients; designers; planning supervisors (CDM co-ordinators); principal contractors; contractors; employees and the self-employed'.*

It further notes that everyone involved in the construction process must give adequate regard to health and safety:

'For manual handling this is especially appropriate for designers when considering specification of materials.'

Designers are reminded that where manual handling cannot be avoided there is the need to include features in their design to facilitate the management of manual handling, for example:

- specifying alternative lower weights
- using glass-fibre reinforced plastic (GRP) ornamental features
- allowing enough space for access with mechanical handling equipment
- designing lifting points
- ensuring that design documentation is clearly marked with the weights of materials or components.

[2] *Backs for the Future*, Safe manual handling in construction (HSG 149), HSE (2000).

3.6 The Confined Spaces Regulations 1997

Under The Confined Spaces Regulations 1997, a confined space is defined as:

'any place, including any chamber, tank, vat, silo, pit, trench, pipe, sewer, flue, well or other similar space in which, by virtue of its enclosed nature, there arises a reasonably foreseeable specified risk'.

Such a space has two defining features: a place which is substantially enclosed (although not always entirely) and where there is a serious risk of injury from hazardous substances within the space or nearby. Enclosed spaces not only include closed tanks, vessels and sewers but also inadequately ventilated rooms and silos.

There are issues in respect of:

- flammable substances and oxygen enrichment
- toxic gas, fume and vapours
- oxygen deficiency
- the ingress or presence of liquids
- solid materials that can flow
- the presence of excessive heat.

These Regulations apply to all premises and work situations subject to the HSAW Act with the exception of diving operations and those conducted below ground in a mine.

All designers should note that paragraph 34 of the associated ACoP (L101) explains that:

'Engineers, architects, contractors and others who design, construct or modify buildings structures etc, should aim to eliminate or minimise the need to enter a confined space'.

3.7 The Work at Height Regulations 2005

The Work at Height Regulations 2005 naturally embrace issues directly associated with the highest category of fatalities in the construction workplace.

Consideration should be given to the definition of 'work at height', which means:

(a) work in any place, including a place at or below ground level;
(b) obtaining access to or egress from such a place while at work, except by a staircase in a permanent workplace,

where, if measures required by these Regulations were not taken, a person could fall a distance liable to cause personal injury.'

These Regulations cover the need to avoid risks from working at height, selection of such work equipment, requirements for particular work equipment, fragile surfaces, falling objects, danger areas and inspections.

Since health and safety statutory duties apply to all those who have a duty to influence the process, these Regulations, like many others, involve contributions by designers as well as contractors and clients, particularly, but not exclusively, in respect of maintenance methodology.

3.8 The Control of Vibration at Work Regulations 2005

The Control of Vibration at Work Regulations 2005 have effect with a view to protecting persons against risk to their health and safety arising from exposure to vibration at work. As well as the obligation on the employer to undertake a risk assessment, Regulation 6 requires that:

> *'(1) The employer shall ensure that risk from the exposure of his employees to vibration is either eliminated at source or, where this is not reasonably practicable, reduced to as low a level as is reasonably practicable.*
>
> *(2) Where it is not reasonably practicable to eliminate risk at source pursuant to paragraph (1) and an exposure action value is likely to be reached or exceeded, the employer shall reduce exposure to as low a level as is reasonably practicable by establishing and implementing a programme of organizational and technical measures which is appropriate to the activity.'*

These measures include elimination, reduction, ergonomic design, provision of auxiliary equipment, appropriate maintenance programmes, design and layout of workplaces, suitable and sufficient information, limitation of exposure, appropriate work schedules and provision of protective clothing.

They are therefore of relevance to designers in consideration of site procedures, equipment/plant choice, logistical considerations and provision of maintenance methodologies.

3.9 The Control of Noise at Work Regulations 2005

The duties arising from the Control of Noise at Work Regulations 2005 are in addition to the general duties set out in the HSAW Act 1974. Damage to hearing from noise is both accumulative and irreversible and there are issues in respect of direct exposure to noise during the initial survey process as well as equipment and machinery design and associated processes of construction.

For those designing equipment and machinery, paragraph 54 of L108 reminds us that:

> *'... machinery is designed and constructed to reduce risks from noise to the lowest level taking account of technical progress...In deciding whether you have done enough to reduce the risks from noise you will need to consider whether there are lower-noise alternatives to the tools and machinery you are using.'*

The duty holder should also note that:

> *'Hearing protection should not be used as an alternative to controlling noise by technical and organisational means, but it can be used as an interim measure while these other controls are being developed.'*

Designers in these areas are also key to the provision of relevant information.

3.10 The Control of Substances Hazardous to Health Regulations (amended 2005)

Regulation 11(6) of the CDM Regulations 2007 requires that:

> *'The designer shall take all reasonable steps to provide with his design sufficient information about all aspects of the design of the structure or its construction or maintenance as will adequately assist—*
>
> - *clients;*
> - *other designers; and*
> - *contractors,*
>
> *to comply with their duties under these Regulations'.*

The designer should be mindful of this requirement in the need to provide COSHH-related information by way of inputs to both pre-construction information and the health and safety file. The COSHH risk assessment enables a valid decision to be made about the measures necessary to prevent or adequately control the exposure of employees to substances hazardous to health arising from the work. Designers should therefore note that it is the provision of residual hazard information, such as the corresponding material hazard data sheets, that is the essential information to be provided.

The competence of all construction professionals is judged not on what is known but on what should be known by virtue of the function fulfilled. It is therefore essential within the design team that the level of health and safety awareness is appropriate to the project content and the duty holder roles to be fulfilled. This does not equate to the designer becoming a health and safety professional, but a competent approach demands that the health and safety professional is embraced by the design team compatible with the complexity of the project issues.

One measure of competence is knowing when the team does not have the skills to fully discharge their duties and that team output requires the input of those with more detailed knowledge in associated areas.

Section 4
THE CDM PROCESS

The holistic approach to construction is illustrated in Figure 4.1 (based on a conventional form of procurement) and includes all the stages from concept/feasibility, through the intervening stages, up to and including the point of obsolescence, which embraces aspects of demolition and decommissioning (see also Plate 1 in the colour plate section).

The CDM Regulations 2007 place statutory duties on various parties that have influence and responsibility during these stages and the following comments serve to annotate the features of Figure 4.1.

A design-and-build procurement strategy would naturally reflect an overlap between the design and construction phases as shown on the central diameter of the diagram.

As indicated, on every construction project there will usually be a client, designer(s) and contractor(s) and, as such, these parties incur statutory duties regardless of the cost or duration of the project. Once the project becomes a notifiable project (longer than 30 days of construction time or greater than 500 person days of construction time) then such a project involves further statutory duty holders, namely the CDM co-ordinator and the principal contractor. Both these appointments must be made in writing by the client (Regulation 14(1) and 14(2)) and become synonomous with the notifiable project.

4.1 Timing

Timing is critical, with the CDM co-ordinator appointed as soon as practicable after 'initial design' and the principal contractor appointed 'as soon as is practicable after the client knows enough about the project to be able to select a suitable person for such appointment'. It should be noted therefore that design has already started before a CDM co-ordinator has to be appointed, but should not have progressed beyond 'initial design'.

There is much merit in early contractor involvement in order to fully appreciate the health and safety management issues and to gain the necessary perspective on the associated concepts of constructability implicit in all the stages of the project. The earlier this focus is gained by the project team, the greater the opportunity to deal with related issues in a viable context. Whilst various forms of procurement lend themselves to this approach, in the conventional admeasurement form of procurement the contractual appointment of the principal contractor takes place much later. However the design team should strive for full constructability appreciation at the earliest possible stage regardless of procurement strategies.

Contractors, facility managers, maintenance personnel and component replacement teams can all offer invaluable insights to this process. These insights should be exploited in order to deliver successful projects.

4.2 Pre-construction information

Pre-construction information is associated with all projects and should represent an information stream towards which all duty holders contribute and from which they all receive relevant information at the appropriate time. It is triggered by the client on all projects (non-notifiable and notifiable) but comes under the management of the CDM co-ordinator on notifiable projects. For this purpose, Regulation 10 requires the client to provide such information to all those designing the structure and to those contractors who have or will be appointed directly by the client. The diagram also indicates that such information could go out to work-package designers, who are appointed well into the construction phase.

Figure 4.1

HOLISTIC DIAGRAM OF THE CONSTRUCTION PROCESS

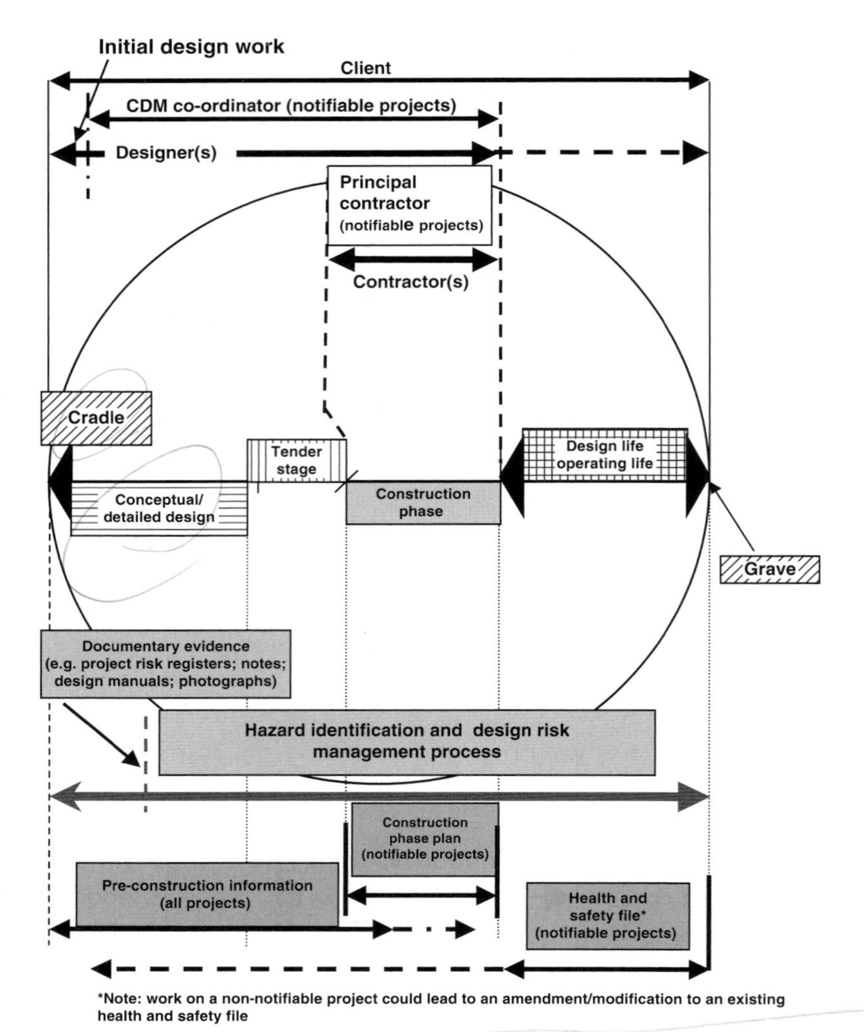

Figure 4.1 Holistic diagram of the construction process. See also Plate 1 in the colour plate section.

Thus all designers must receive relevant health and safety-related information to enable them to fulfil their role on all projects. The position of 'lead designer' can greatly facilitate the co-ordination of this process.

4.3 Construction phase plan

The construction phase plan exists only on notifiable projects and is the articulation of control details and contingency measures for the health and safety management of the construction process. These measures are exercised by the principal contractor throughout the project. Pre-construction information feeds into this process both before the start of the construction phase and thereafter. Particular control needs to be exercised by the client and CDM co-ordinator in determining the sufficiency of this document for its stated purpose of achieving a 'safe and suitable' working environment from the first day of the construction phase. Thereafter, it becomes the principal contractor's duty to ensure its further development in a manner compatible with the programme of works and subject to the client's duty to ensure the ongoing effectiveness of management arrangements throughout the project (Regulation 9(2)). The CDM co-ordinator is instrumental in looking after the client's interests in this respect.

Designers need to be aware that late design changes often have an impact on the development of the construction phase plan. The interface between designers, principal contractor and the CDM co-ordinator all must provide focus on the compatibility of the construction phase plan and its development in respect of variation orders and architects' instructions.

Additionally, late design change for whatever reason, carries a disproportionate cost element because it has to be accommodated within the developed boundaries of completed or almost completed designs. Clients often fail to appreciate these implications and their further impact on health and safety management. Designers and others must continue to inform and educate lay clients.

4.4 Health and safety file

The health and safety file, as the Approved Code of Practice (ACoP) notes, '*should contain the information needed to allow future construction work, including cleaning, maintenance, alterations, refurbishment and demolition to be carried out safely*'. On a new notifiable project, the health and safety file is handed over to the client by the CDM co-ordinator at the end of the construction phase as either a new document or a revision/amendment to an existing document. The client's management systems should also cater for amendments to be made to existing health and safety files where changes arise out of work of a non-notifiable nature. This latter situation should be a perspective focused on by designers since, as well as the customary duty holder role, they also inevitably function as the professional adviser to the client and subsequently need to remind him of his duty under Regulation 17(3)(b) whereby:

> *'The client shall take reasonable steps to ensure that after the construction phase the information in the health and safety file is revised as often as may be appropriate to incorporate any relevant new information.'*

An existing health and safety file remains a primary source document for any desktop study in the early stages of all projects. After having its existence confirmed, the designer should then ensure that it contains current information.

The solid (red) line in Figure 4.1 is a 'drop-down' from the central diameter and provides a perspective on the time horizon to be embraced by the design teams. The design (health and safety) risk management process must give due consideration to all the health and safety issues forseeably arising during any one of the phases of the project up to and including demolition/decommissioning. This statutory requirement does not extend to second-guessing, but it must ensure that issues are accounted for, contributions made and residual issues communicated within a framework of co-ordination that embraces all relevant design teams. This communication aspect of the design process must focus on residual issues significant (and principal) to the specifics of the project. This cannot be done if the design team members' level of awareness is inappropriate to the discharge of their duties. This issue itself touches on the competence of the team.

It is essential that the design risk management process is demonstrable without being enslaved to paperwork systems that add very little to the effectiveness of communication. This will always be a challenge, but the designer is entitled to design on the assumption that work is to be undertaken by a competent contractor. Information from the design process into pre-construction information and the health and safety file does not therefore need to focus on issues that would normally be expected to be dealt with by the contractor and/or facility management team.

For demonstrable evidence of the design process refer to Chapter 7.

Communication systems therefore need to be focused on the passage of succinct and relevant information that is specific to the project. Figure 4.2 offers a systems approach to the design process, with inputs to the process coming from the list on the left and outputs from the design process forming the list on the right.

Collection and *collation* can all be seen as constituting pre-construction information, amassed initially from the client' s archives and supplemented by the design desktop study and further enquiries.

Consideration, co-ordination, contribution and communication represent the discharge of design duties under Regulations 11 and 18, and on the medium-to-large project would be delivered by the multi-disciplinary design team under the co-ordination of the 'lead designer'. As indicated on Figure 4.1 this could be evidenced by project risk (health and safety) registers, notes, design manuals, photographs and so on.

Adequate information for work to be undertaken safely represents design duties arising from Regulations 11(6) and 18(2) and can take the form of any of the items listed to the right of Figure 4.2. In design process terms it conveys information on residual hazards arising from the duty to eliminate and reduce hazards, compatible with the term 'as far as reasonably practicable'.

Other parties besides designers manage health and safety risks associated with construction. This arises from the basic premise that risk should be managed by the party best positioned to manage it. However, it does not translate into contractors having to manage risks arising from an ineffective design process. Such an outcome would represent non-compliance with a statutory duty and remains unacceptable.

Table 4.1 (page 56) identifies relevant regulations affecting all duty holders.

The details of such regulations are provided in Table 4.2 (pages 58 and 59) for all duty holders with the more specific designer-related regulations provided in Tables 5.1a (page 70) and Table 5.1b (page 72).

Figure 4.2
SYSTEMS APPROACH

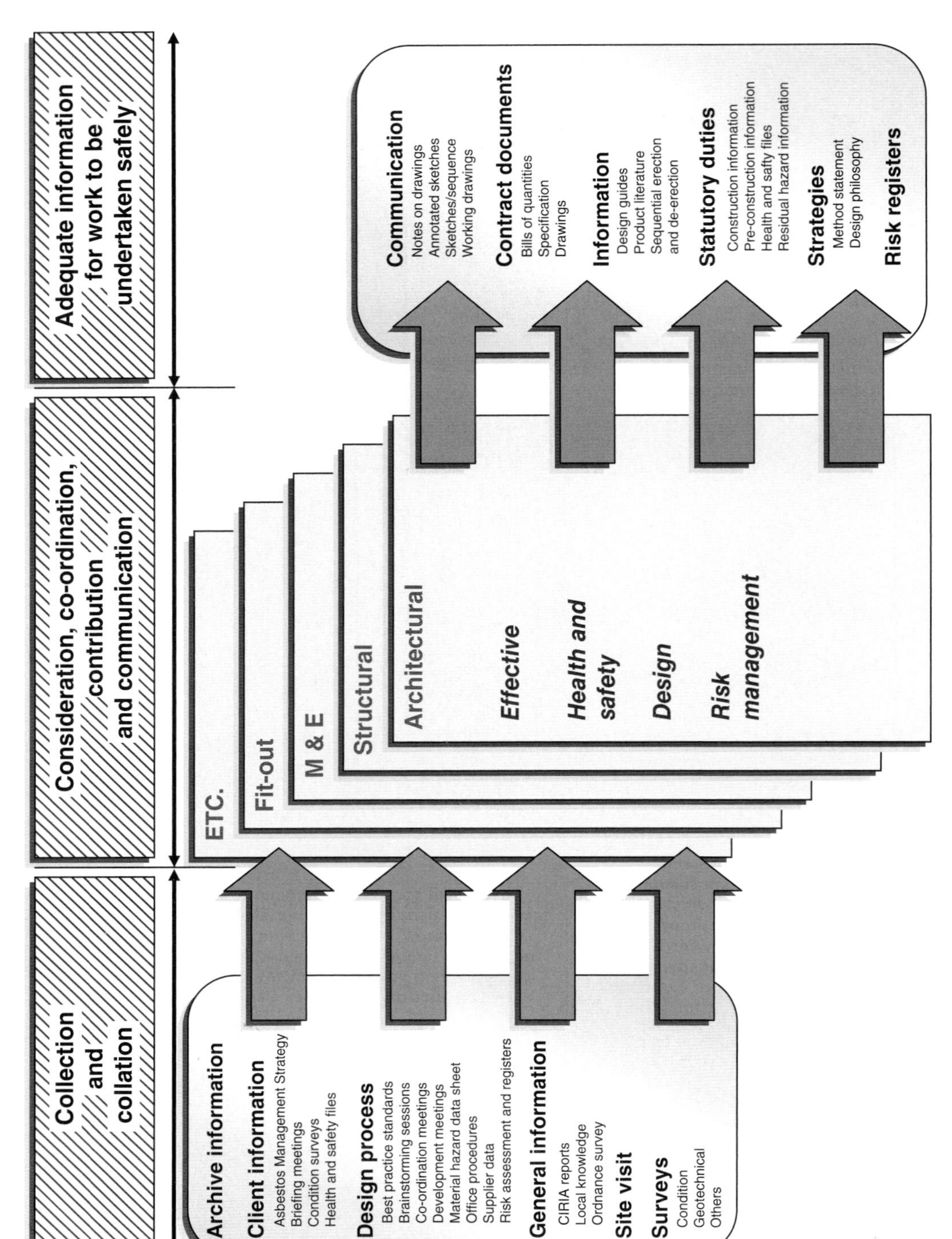

Figure 4.2 Systems approach.

Table 4.1

APPLICABLE REGULATIONS FOR DUTY HOLDER COMPLIANCE

	Section	Part 2										Part 3											Part 4	
	Regulation	4	5	6	7	8	9	10	11	12	13	14	15	16	17	18	19	20	21	22	23	24	25(2)	26 to 44
Duty holder																								
Client		x	x	x	x	x	x	x		x		x	x	x	x								x	(x)*
Designer		**x**	**x**	**x**	**x**				**x**	**x**						**x**							**x**	**x***
Contractor		x	x	x	x					x	x						x						x	x
CDM co-ordinator		x	x	x	x													x	x				x	(x)*
Principal contractor		x	x	x	x					x	x						x			x	x	x	x	x

The CDM co-ordinator and principal contractor are only appointed on the notifiable project.
*Apply only if Regulation 25(2) applies. Unshaded Regulations apply to all construction projects, shaded Regulations additionally apply to notifiable projects. Part 1 (Regulations 1–3) deals with interpretations and application and is not included. Part 5 (Regulations 45 to 48) deals with 'General civil liability, enforcement in respect of fire; transitional provisions and revocations and amendments' and is not included.

Table 4.2

CONSTRUCTION (DESIGN AND MANAGEMENT) REGULATIONS 2007

<table>
<tr><th>Duty holder</th><th>ALL DUTY HOLDERS</th><th>ALL PROJECTS</th></tr>
<tr><th>Regⁿ</th><th>Description</th><th>Comment</th></tr>
<tr><td>4.</td><td>Competence
1) No person on whom these Regulations place a duty shall:
a) appoint or engage a CDM Co-ordinator, designer, principal contractor or contractor unless he has taken reasonable steps to ensure that the person to be appointed or engaged is competent;
b) accept such an appointment or engagement unless he is competent;
c) arrange for or instruct a worker to carry out or manage design or construction work unless the worker is:
(i) competent; or
(ii) under the supervision of a competent person.</td><td>The competence issue extends to all those who appoint and are appointed, not just the client as a duty-holder.
• It must be demonstrable that reasonable steps have been taken to establish competence.
• In accepting the appointment, the party self-certifies their competence to perform the duties.
The issue of competence also relates to arrangements for the management of design or construction work being undertaken by a competent person or under the supervision of a competent person. As previously, this competence extends only to the discharge of their health and safety duties.
This has even greater importance for companies, contractors, CDM co-ordinators and designers by virtue of the content of Appendices 4 and 5 of the ACoP, which highlights aspects of competence and gives guidance for assessing it, both at corporate and individual level.
As an issue, competence resonates with the contents of Research Report 422 and the registers of CDM co-ordinators and designers being launched by APS, ICE, RIBA and others.</td></tr>
<tr><td></td><td>2) Any reference in this Regulation to a person being competent shall extend only to his being competent to:
a) perform any requirement; and
b) avoid contravening any prohibition,
imposed on him by or under any of the relevant statutory provisions.</td><td>Competence should be assessed prior to appointment.
Appendices 4 and 5 of the ACoP provide guidance on the establishment of competence for all the appointed duty-holders.</td></tr>
<tr><td>5.</td><td>Co-operation
(1) Every person concerned in a project on whom a duty is placed by these Regulations, including paragraph (2), shall:
a) seek the co-operation of any other person concerned in any project involving construction work at the same or an adjoining site so far as is necessary to enable himself to perform any duty or function under these Regulations; and
b) co-operate with any other person concerned in any project involving construction work at the same or an adjoining site so far as is necessary to enable that person to perform any duty or function under these Regulations.
(2) Every person concerned in a project who is working under the control of another person shall report to that person anything which he is aware is likely to endanger the health or safety of himself or others.</td><td>This emphasises the need for all duty holders to co-operate not only with those on their own construction project but with those on adjoining sites, who are involved in construction work. This acknowledges the impact on the project of what is happening adjacent to it and the influence such activities might have on the health and safety management interfaces between projects.
There is a need for all duty holders to proactively seek the co-operation of those parties. Likewise they must actively co-operate with duty holders from those adjoining projects.
An awareness of health and safety endangerment should also lead to reports to those in control so that deficiencies in safe and suitable systems of work can be remedied.
The adjoining site interpretation should not be taken too locally.</td></tr>
</table>

<table>
<tr><th>Duty holder</th><th>ALL DUTY HOLDERS</th><th>ALL PROJECTS</th></tr>
<tr><th>Regⁿ</th><th>Description</th><th>Comment</th></tr>
<tr><td>6.</td><td>Co-ordination

All persons concerned in a project on whom a duty is placed by these Regulations shall co-ordinate their activities with one another in a manner which ensures, so far as is reasonably practicable, the health and safety of persons:

a) carrying out the construction work; and
b) affected by the construction work.</td><td>All duty holders to ensure works are co-ordinated and effectively managed to ensure that health and safety is not jeopardised.

The ACoP introduces the concept of the 'lead designer' (para 48) for design co-ordination, but this Regulation encompasses more than the design process and relates to the co-ordination of all construction-related activities.

Para 49 also identifies that for the notifiable project, CDM co-ordination is more than simply design co-ordination – hence care should be taken. This Regulation therefore has an impact on planning, programming and general management.

Central to the theme of co-ordination is paragraph 20 of the ACoP, which provides the following guidance:

'The architect, lead designer or contractor who is carrying out the bulk of the design work should normally co-ordinate the health and safety aspects of the design work; the builder or main contractor, if there is one, should normally co-ordinate construction work.'

As noted by the Construction Industry Advisory Committee working group, early-stage contractor co-ordination over site-wide issues such as location of services, structures and feasibility are known to provide significant opportunities for effective elimination or avoidance of major hazard issues.

Design co-ordination is also critical to the iterative nature of the design process.

Both the above perspectives have implications for the duty holder on both the non-notifiable and the notifiable project.</td></tr>
<tr><td colspan="3">Regulation 5 (co-operation) and Regulation 6 (co-ordination) have implications throughout, on the management arrangements of all duty holders and in particular those of the client, by virtue of duties in respect of Regulation 9 (Clients' duties). (Refer to Section 6.1, pages 81 and 82)</td></tr>
<tr><td>7.</td><td>General principles of prevention

(1) Every person on whom a duty is placed by these Regulations in relation to the design, planning and preparation of a project shall take account of the general principles of prevention in the performance of those duties during all the stages of the project.
(2) Every person on whom a duty is placed by these Regulations in relation to the construction phase of a project shall ensure so far as is reasonably practicable that the general principles of prevention are applied in the carrying out of the construction work.</td><td>This re-states the general tenets of health and safety management introduced in Schedule 1 of the Management of Health and Safety at Work Regulations 1999.Also now covered in Appendix 7 of the ACoP.

A hazard should be removed so far as is reasonably practicable so that others lower down the supply chain do not have to manage it.

The principles consist of:

• avoiding
• evaluating
• combating at source
• adapting the work to the individual
• adapting to technical progress
• replacing
• prevention policy
• collective protection measures
• appropriate instructions

Note the emphasis on all stages of the project.

These are the basic tenets of health and safety management first outlined in Schedule 1 of the Management (Health, Safety and Work) Regulations 1999 (formerly 1992) and now repeated in Appendix 7 of the ACoP.

Compliance with the detail outlined in Schedule 1 dictates how designers and contractors should have been approaching their management role, certainly since 1 January 1993.</td></tr>
</table>

The above duties apply to all duty holders on all projects. For additional and specific design duties, refer to pages 70 and 72.

Section 5
ROLE OF THE DESIGNER

The design team occupies a unique position in the construction process, not only through design per se but also through the associated role of professional adviser to the client. The HSE acknowledge this and continue to emphasise the influential position of design in the delivery of effective health and safety management.

As illustrated in Figure 4.1, the designer's perspective takes in all those foreseeable aspects of health and safety associated with constructability, useability, maintainability and replacement up to and including demolition and dismantling. The design team must implement general principles of prevention so that others lower down the supply chain will only have to manage residual hazards that have not been eliminated.

Fundamentally,

'Designers must not produce designs that cannot be constructed, used, maintained, replaced and removed safely.'

This is facilitated by a full understanding of the detail and philosophy associated with the CDM Regulations 2007. Designers must note the extension to the definition of design outlined in Regulation 2(1), which now includes:

- drawings
- design details
- specification and bills of quantities, including specification of articles or substances relating to a structure
- calculations prepared for the purpose of design

and the need in designing a structure as a place of work to account for the provisions of the Workplace (Health, Safety and Welfare) Regulations 1992.

The word 'designer' as used in Regulation 2(1) refers to:

'... any person (including a client, contractor or other person referred to in these Regulations) who in the course or furtherance of a business:

(a) prepares or modifies a design; or
(b) arranges for or instructs any person under his control to do so

relating to a structure or to a product or mechanical or electrical system intended for a particular structure. A person is deemed to prepare a design where a design is prepared by a person under his control.'

Many parties inadvertently take on design duties through design involvement; specification influence; design alteration; imposition of a particular method of work or material, either in default or direct mode. Further guidance is given in paragraph 116 of the Approved Code of Practice (ACoP).

5.1 Who are designers?

They are:

- Anyone who specifies or alters a design
- Anyone who stipulates a particular method of work or material

- Architects
- Building services
  - electrical
  - heating and ventilating
  - mechanical
  - testing and commissioning
- Consultants generally, including those involved with:
  - civil engineering
  - drainage
  - ecology
  - electrical aspects
  - environment
  - foundation issues
  - geotechnics
  - highway and traffic engineering
  - hydrology
  - interior design
  - landscape
  - mechanical issues
  - piling
  - structural, i.e for concrete; masonry; steelwork; timber
  - telecommunications
  - traffic
  - transportation
  - waste management
- Designers of fixed plant
- Principal contractors who manage design and build
- Project managers who influence change or methods of working
- Specialist work package contractors (design and build), i.e cladding, fit-out, glazing; windows
- Specifiers, i.e. buyer, client, contractor, point of contact salesperson
- Temporary works designers.

However, manufacturers supplying standardised components have no duties as designers under the CDM Regulations 2007, provided they are not modified as a bespoke product (refer to paragraph 118 of ACoP).

The range of designer duties is set out in Table 4.1. In addition to the general duties under Regulations 4, 5, 6 and 7, more specific duties reside with the design team on both notifiable and non-notifiable projects under Regulations 11 and 12 (refer to Table 5.1a). There are then further duties under Regulation 18 on notifiable projects (refer to Table 5.1b).

These are all captured in Figure 5.1.

These duties impose action points for the designer (Regulations 11(1) and 18(1)) in regulating the client's involvement with the process of construction, as well as in imposing process controls on the management of the design process itself.

For the design team there is less of a cultural journey involved in achieving compliance with these Regulations compared with that demanded of the client. This is simply because a compliant design process under the 1994 Regulations should have substantially laid the foundations for meeting the requirements of the 2007 Regulations.

Figure 5.1
THE DESIGNER'S DUTIES

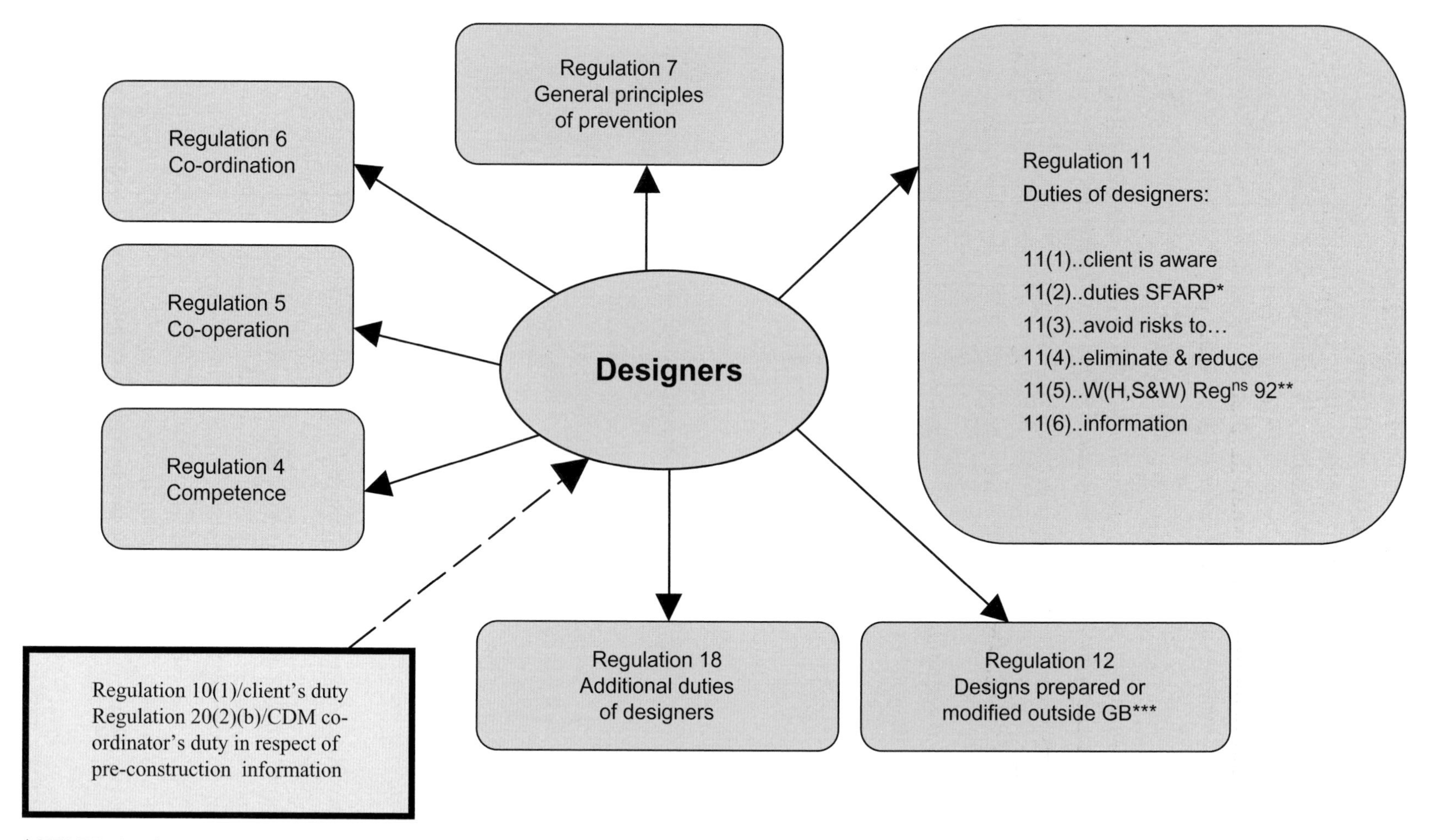

*SFARP, 'so far as is reasonably practicable'
**W(H,S & W)Regns, Workplace(Health, Safety and Welfare) Regulations 1992
***GB, Great Britain and its territorial waters (can also include Northern Ireland)

Figure 5.1 The designer's duties.

Whilst the designer occupies a central position in health and safety management within the project process, he is also only one of three or five duty holders, depending on the non-notifiable or notifiable status of the project. As such, the design team is not omnipotent, but simply a vital piece of the project jigsaw.

An accident on a construction site will not rebound to haunt the designer provided the design team can demonstrate that they have discharged their duties 'so far as is reasonably practicable'. Other duty holders must inevitably play their part.

Nonetheless, the effective CDM model implicitly demands that all duty holders address and own the issues that they are best placed to manage. This requires ownership and awareness of associated health and safety issues based on a contribution around the general principles of prevention and culminating in communication to others so that they also can discharge their statutory duties.

It is a process that is based on the concept of the integrated team and which aims to be a seamless, joined-up-thinking process with responsibility for health and safety management demonstrated by all those in a position of influence. The design team is obviously involved but not exclusively.

The design management process, like project management, can fail through ineffective communication, lack of ownership, team fragmentation, resource limitation, incompetence, inappropriate levels of awareness, risk management failure, complacency, reactive responses, infrastructure deficiencies, confusion and unrealistic timescales.

Table 5.1a
DESIGNER DUTIES (ALL PROJECTS)

Duty holder	DESIGNER	ALL PROJECTS
	Description	**Comment**
Reg^{n}	***NOTE: Regulations 4, 5, 6 and 7 apply to ALL DUTY HOLDERS CONCERNED IN A PROJECT (Refer to pages 58 and 59)***	
11.	**Duties of designers** (1) No designer shall commence work in relation to a project unless any client for the project is aware of his duties under these Regulations. (2) The duties in paragraphs (3) and (4) shall be performed so far as is reasonably practicable, taking due account of other relevant design considerations. (3) Every designer shall in preparing or modifying a design which may be used in construction work in Great Britain avoid foreseeable risks to the health and safety of any person: (a) carrying out construction work; (b) liable to be affected by such construction work; (c) cleaning any window or any transparent or translucent wall, ceiling or roof in or on a structure; (d) maintaining the permanent fixtures and fittings of a structure; or (e) using a structure designed as a workplace. (4) In discharging the duty in paragraph (3), the designer shall: (a) eliminate hazards which may give rise to risks; and (b) reduce risks from any remaining hazards and in so doing shall give collective measures priority over individual measures. (5) In designing any structure for use as a workplace the designer shall take account of the provisions of the Workplace (Health, Safety and Welfare) Regulations 1992 which relate to the design of, and materials used in, the structure. (6) The designer shall take all reasonable steps to provide with his design sufficient information about aspects of the design of the structure or its construction or maintenance as will adequately assist: (a) clients; (b) other designers; and (c) contractors to comply with their duties under these Regulations.	These duties build on the influential position of the designer in respect of health and safety management within the supply chain. Revisit Regulation 2 and the interpretation of design, which now includes: • drawings • design details • specification and bill of quantities as well as calculations prepared for the purpose of design and the need to: • design for use as a place of work. These duties apply to *all construction projects* Subsequently the health and safety management strategies (Regulation 7) and supporting communication systems attached to the design process need to deliver to both the informed and lay client. Note: 1. Need to ensure all clients are aware of their duties before commencing work. This now also includes public sector clients. (11(1)) 2. Emphasis on the 'quantum of risk' approach, balancing up the consideration of hazards in one scale and the design issues of form, function, fitness of purpose, environmental impact, cost, health and safety and other construction-related legislation in the other. All with regard to the qualifying term 'so far as is reasonably practicable'. (11(2)) 3. Key words here are 'preparing and modifying', with particular mention of 'maintenance' and 'using a structure designed as a workplace'. (11(3)) 4. Important to read in conjunction with the AcoP since the elimination of hazards is to be discharged 'so far as is reasonably practicable'. (Regulation 11(2) and ACoP para.125) 5. Fixed workplaces such as factories, offices, schools and hospitals must account for the provisions of the Workplace (Health, Safety and Welfare) Regulations 1992 relating to the design of, or materials used in the structure. 6. Much greater emphasis on the holistic project management process, with designers having to provide relevant information to numerous other duty holders. Information referred to should be significant/principal and project-specific.
12.	**Designs prepared or modified outside Great Britain** Where a design is prepared or modified outside Great Britain for use in construction work to which these Regulations apply: (a) the person who commissions it, if he is established within Great Britain; or (b) if that person is not so established, any client for the project shall ensure that Regulation 11 is complied with.	Builds on the Paul Wurth case* and the CDM Regulations 2000 amendment. This has potential impacts for designer; contractor and obviously the client.

*Regina *v.* Paul Wurth SA, Court of Appeal (26 January 2000).

Table 5.1b

DESIGNER DUTIES (ADDITIONAL DUTIES ON NOTIFIABLE PROJECTS)

Duty holder	DESIGNER (Additional duties where project is notifiable)	
Reg^n	**Description**	**Comment**
18.	**Additional duties of designers** (1) Where a project is notifiable, no designer shall commence work (other than initial design work) in relation to the project unless a CDM co-ordinator has been appointed for the project. (2) The designer shall take all reasonable steps to provide with his design sufficient information about aspects of the design of the structure or its construction or maintenance as will adequately assist the CDM co-ordinator to comply with his duties under these Regulations, including his duties in relation to the health and safety file.	• For notifiable projects all designers must ensure that a CDM co-ordinator has been appointed immediately the stage of 'initial design' work has been passed. • What is 'initial design' work? There is a distinction within the ACoP between 'initial design' and 'significant design'. • The latter includes 'preparation of initial concept design' and 'implementation of any strategic brief'. (para 66 of the ACoP) • Note: Such appointments must be in writing. • Significant/principal hazard information in respect of the future operability and usability of the structure, up to and including the final demolition/ decommissioning process, must be provided to the CDM co-ordinator by the design teams for inclusion in the health and safety file. • This, as ever, is over and above day-to-day preventative maintenance information. If a CDM co-ordinator hasn't been appointed at the end of the 'initial design stage' then by default the client takes on the duty. However the design process should discontinue until such an appointment has been made. (Regulation 14(4))

As outlined in paragraph 119 of the ACoP designers should, *for all projects*:

- ensure they are competent and adequately resourced
- co-operate with other duty holders on this and adjoining sites
- co-ordinate their activities to improve the management/control of risk
- design in conjunction with the general principles of prevention
- ensure client is aware of his or her duties
- account for the provisions of the Workplace (Health, Safety and Welfare) Regulations 1992
- provide sufficient (and suitable) information about aspects of design to:
    - clients
    - other designers
    - and contractors.

Co-operation and co-ordination must acknowledge the working practices of industry and extend to the relevant sub-elements that are an intrinsic part of most projects, namely sub-design, specialist work packages and temporary design.

Additionally on a *notifiable* project, designers must:

- ensure the client has appointed a CDM co-ordinator in writing before embarking on '*significant detailed design*'
- provide relevant information to the CDM co-ordinator.

As with all other duty holders, focus cannot solely be on the specifics of the CDM Regulations 2007, for it is the whole portfolio of construction-related legislation that is relevant to the task that needs to be considered, including Regulation 3 of Management of Health and Safety at Work Regulations 1999, which broadly requires all employers:

- to undertake an assessment of the risks to the health and safety of their employees and to other persons arising out of, or in connection with, their work
- to make appropriate arrangements for implementing any preventative/protective methods identified in the risk assessment.

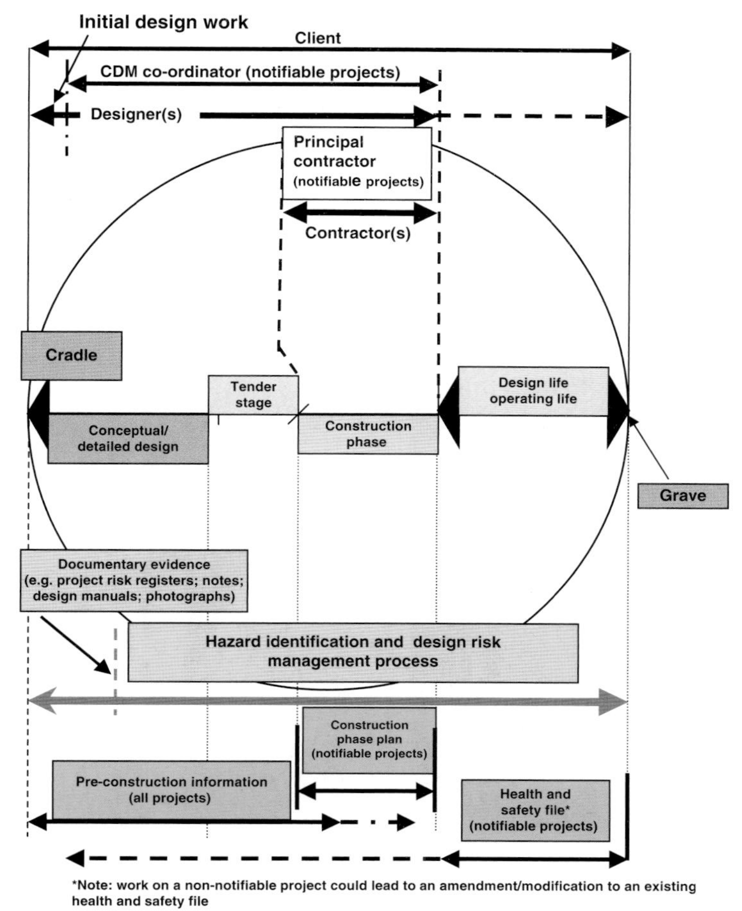

Plate 1 Holistic diagram of the construction process.

Designer duties		Interfaces with other duty holders				
Duty holder	Designers' duties	Client	Other designers	CDM co-ordinator	Principal contractor	Contractor
Duties in this row may establish interfaces with the duty holder and the designer	Regulations	9 10(1)		20(2);(b) 20(2)(c) 20(2)(d) 22(1)(b)*	13(2) 22(1)(a) 22(1)(b)*	13(2)
Designer	4	Competence	Competence		Competence	Competence
	5		Co-operation	Co-operation	Co-operation	Co-operation
	6		Co-ordination	Co-ordination	Co-ordination	Co-ordination
	7					
	11(1)	Client awareness				
	11(2)					
	11(3)					
	11(4)					
	11(5)					
	11(6)	Information	Information		Information	Information
	12		Design outside GB			
	18(1)	CDM co-ordinator appt.				
	18(2)			Information		
	25(2)**				Construction work	Construction work

Note: * Regulation 22(1)(b) requires liaison between the principal contractor and the CDM co-ordinator in respect of design or changes to a design
** Regulation 25(2) reminds, that any person other than a contractor who controls the way in which a contractor has to work could incur duties in respect of the management of the construction work (Regulations 26 to 44). In other words it is imperative that the designer does not impose construction methodology onto the contractor unless dictated by operational necessity

Plate 2 Design interface with other duty holders.

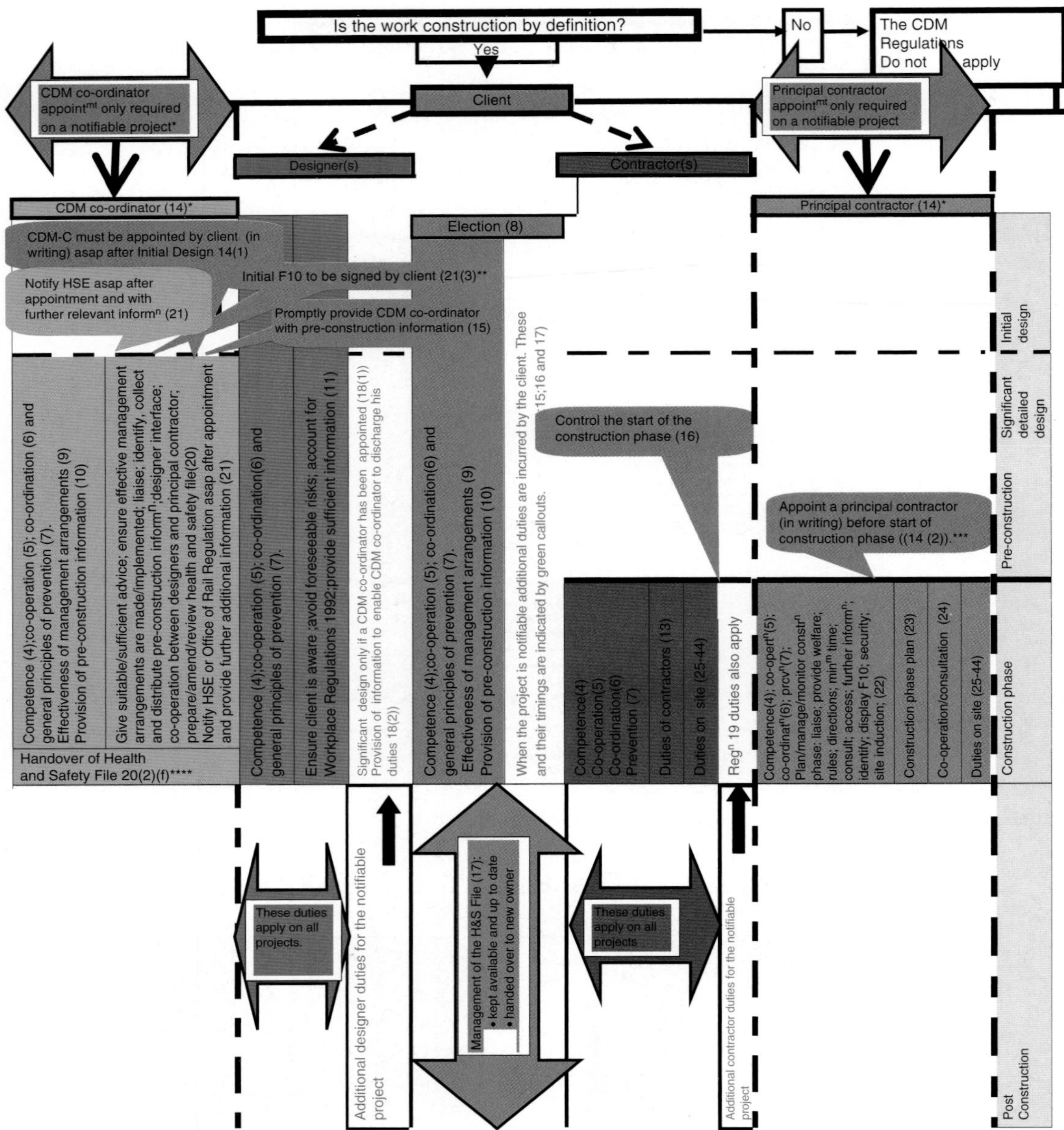

NOTES:

*Construction projects become notifiable if the construction phase exceeds 30 days or more than 500 person days. (refer Regulation 2(3)).

** Notification to the HSE/Office of Rail Regulation to be signed by client or someone on his behalf (Schedule 1)

*** It is recommended (regulation 14(2)) that the principal contractor is appointed as soon as practicable after the client knows enough about the project to be able to select a suitable person (i.e it depends on the procurement strategy)

**** Health and safety file to be handed over by CDM co-ordinator at end of the construction phase

Legend:

Client	Duties to be undertaken by clients on all projects
Client	Additional duties to be undertaken by client on notifiable projects
Control the start of the construction	Additional duties and timings to be undertaken by client
Designer	Duties to be undertaken by designers on all projects
Designer	Additional duties to be undertaken on notifiable projects
Contractor	Duties to be undertaken by contractors on all projects
Contractor	Additional duties to be undertaken on notifiable projects
CDM co-ordinator	Duties to be undertaken by CDM co-ordinator on notifiable projects
Principal contractor	Duties to be undertaken oby principal contractor on notifiable projects

S.D.Summerhayes/June 2009

Plate 3 Roadmap.

Section 6

THE DESIGN RISK MANAGEMENT PROCESS

Design risk management is an intrinsic part of the design process regardless of the project status, i.e. whether notifiable or not. Its generic embrace would cover all risks (political, environmental, social and technical), but for our purposes must focus on relevant aspects of health and safety.

In keeping with good project management principles, the sooner an issue is identified, the greater the time period available for contingency management to respond. This depends on a proactive approach, with the design team members conscious throughout that 'radar screens' must be turned on and tuned in to facilitate the effective discharge of statutory duties. It is only via the integrated team approach that project health and safety management can be successfully delivered.

Boundary conditions imposed by the client are an inevitable part of all projects, but designers, like all other duty holders, must remain vigilant and ensure that such restraints, particularly finance and time, never compromise the discharge of statutory duties.

The process of design risk management, as illustrated in Figure 4.1, begins at the conceptual stage and continues throughout all the intervening phases up to and including all relevant aspects of de-commissioning and dismantling. It occurs whenever design takes place, regardless of the existence of a client or the appointment of a CDM co-ordinator.

Paragraphs 124 and 125 of the Approved Code of Practice (ACoP) advise that:

'124. Designers have to weigh many factors as they prepare their designs. Health and safety considerations have to be weighed alongside other considerations, including cost, fitness of purpose, aesthetics, buildability, maintainability and environmental impact. CDM 2007 allows designers to take due account of other relevant design considerations. The Regulations do not prescribe design outcomes, but they do require designers to weigh the various factors and reach reasoned professional decisions.

'125. Designers are required to avoid foreseeable risks "so far as is reasonably practicable", taking due account of other relevant design considerations. The greater the risk, the greater the weight that must be given to eliminating or reducing it. Designers are not expected to consider or address risks which cannot be foreseen, and the Regulations do not require zero risk designs because this is simply impossible. However, designers must not produce designs that cannot be constructed, maintained, used or demolished in reasonable safety.'

Designers therefore have to account for this 'quantum of risk' approach in the knowledge that health and safety is not meant to dominate design but is meant to be fully integrated into the design process, with the design perspective giving consideration to constructability issues without imposing methods on the contractor.

Throughout the Regulations duty holders have to discharge duties:

'so far as is reasonably practicable.'

The Court of Appeal's interpretation of this phrase is:

'Reasonably practicable is a narrower term than physically possible, and implies that a computation must be made in which the quantum of risk is placed in one scale, and the sacrifice, whether in money, time or trouble, involved in the measures necessary to avert the risk, is placed in the other; and that, if it be shown that there is a gross disproportion

between them, the risk being insignificant in relation to the sacrifice, the person upon whom the duty is laid discharges the burden of proving that compliance was not reasonably practicable. This computation falls to be made at a point of time anterior to the happening of the incident complained of.'

Hence the designer is allowed choice, based on his informed professional judgement, and there may be numerous options to be exercised in discharging his duties. The design risk management process does not seek to dictate one avenue of pursuit, but the various options pursued and the solution eventually chosen must provide visibility of the discharge of duties through support provided via suitable transparency and documentary evidence.

The process is therefore subjective, influenced by industry norms, good practice and sector guidance. Designers can undertake the process in different ways, subject to these influences, and there may be many options that can be followed, each one of which represents an approach that is fully compliant with the requirements of the law. For some options, because of the design route taken, there will be a greater need to communicate as a result of the residual hazards that output from the process.

The argument in support of the outcome must be coherent, based on full awareness of health and safety related issues, and demonstrable in the process of *consideration*; *contribution* and *communication*.

As noted in the *Key Points* of CIRIA Report C662:[1]

'A designer's duties under CDM require considerable common sense and openness in order to relate to the other duty holders in a constructive manner. Duties must be carried out in a manner that is proportional to the type of project and the likely level of risk. There may be a number of viable and fully compliant design solutions, but the optimum solution would require less additional communication (Regulation 11(6)), because there are fewer residual issues to be communicated. This is what is being aimed for.'

Generally, the designer must consider constructability issues without imposing methodology on the contractor. Construction requires contractors to use their prerogative to choose their method of working, providing it does not affect the permanent structure. This is the basis of contract, reliant on the finesse and expertise of the contractor and reflective of the professional indemnity insurance cover under which he functions. This consideration also impacts on the relationship between constructability and contractual responsibility.

In certain circumstances, because of operational necessity or technical dependency, there are good reasons to impose methodology but these situations should be kept to the minimum.

Regulation 25(2) serves to further remind that the imposition of control on the contractor by others (not only duty holders) inevitably incurs duty associated with the management of construction work as set out in Part 4 of the CDM Regulations 2007.

Regulation 25(2) ...

'Every person (other than a contractor carrying out construction work) who controls the way in which any construction work is carried out by a person at work shall comply with the requirements of regulations 26 to 44 insofar as they relate to matters which are within his control.'

---

[1] *CDM 2007 Construction Work Sector Guidance for Designers* (3rd edn), CIRIA Report 662, CIRIA (2007).

in a health and safety management process that engages with all design team members, and includes design changes arising from variation orders and architects' instructions, inclusive of temporary works design.

The discharge of duties to Regulations 11(6) and 18(2) depends on the interface established between design teams and the client, other designers and contractors, as well as the CDM co-ordination team.

This infrastructure effectiveness is dependent on the co-operation (Regulation 5) and co-ordination (Regulation 6) established between the various parties. Throughout all stages of a project it remains a legal obligation for the client to ensure that such management arrangements are effectively set up and remain so thereafter (Regulation 9).

Paragraph 48 of the ACoP notes that:

'Nominating one designer as the 'lead designer' is often the best way to ensure co-ordination and co-operation during work which involves a number of designers.'

Such an appointment is essential on the larger project to expedite the process of design risk management, although there is no legality to the term, since it is promoted in the ACoP but not mentioned in the CDM Regulations themselves. However, it provides an invaluable facilitation model for ensuring co-ordination and communication within the process.

(Note: The one common failure mode in project management remains that of communication failure.)

Monitor/review

Project management controls require proactive responses. Procedures depend on individuals and teams for implementation as well as response to changes in information, circumstances and experience. Failure to respond ensures the decline and obsolescence of control systems. The design risk management cycle must therefore feed back as part of the journey of constant improvement. Systems need to be checked for delivery of expected outputs and altered when signals and outcomes differ from those anticipated and intended.

Subject to monitoring, design risk management must continually review its process to avoid bankruptcy and ensure individuals are not rendered vulnerable to unacceptable health and safety outcomes.

As well as the design audit trail contribution there is an advantage in feeding into the wider project culture that underpins organisational competence.

6.1 Additional interfaces

The following interfaces are all relevant to the role of the designer.

Designer/client

Regulation 9 states that:

(1) *Every client shall take reasonable steps to ensure that the arrangements made for managing the project (including the allocation of sufficient time and other resources) by persons with a duty under these regulations (including the client himself) are suitable to ensure that:*

(a) the construction work can be carried out so far as is reasonably practicable without risk to the health and safety of any person;

...

(c) any structure designed for use as a workplace has been designed taking account of the provisions of the Workplace (Health, Safety and Welfare) Regulations 1992 which relate to the design of, and materials used in the structure.

(2) The client shall take reasonable steps to ensure that the arrangements referred to in paragraph (1) are maintained and reviewed throughout the project.

On the notifiable project it would normally be the CDM co-ordination team that would seek confirmation of the continued effectiveness of management arrangements on behalf of the client.

Regulation 10(1)

'Every client shall ensure that:

(a) every person designing the structure ... is promptly provided with pre-construction information ...'

Hence the design teams, under the umbrella of pre-construction information, should promptly receive from the client relevant information:

- about or affecting the site or the construction work
- about the proposed use of the structure
- in any existing health and safety file.

CDM co-ordinator/designer

Regulation 20(2)

'The CDM co-ordinator shall:

...

(b) take all reasonable steps to ensure that designers comply with their duties under Regulations 11 and 18(2)

(c) take all reasonable steps to ensure co-operation between designers and the principal contractor during the construction phase in relation to any design or change to a design.'

Hence the CDM co-ordinator must continually seek assurance that the design risk management process is being duly carried out in an effective and efficient manner.

(Designer)/CDM co-ordinator/principal contractor

Furthermore Regulation 20(1)(c) states that:

'The CDM co-ordinator shall:

Liaise with the principal contractor regarding:

(iii) any design development which may affect planning and management of the construction work'

Design development is a factor of procurement, with more late development associated with certain forms of procurement than others, e.g. 'design and build'. Such development must also accommodate design change, which offers flexibility of response. This, however, should be distinguished from change arising from an inadequately developed brief or an ambiguously drafted scope of works, both of which are unacceptable.

Design change within the confines of a construction programme carries further significance, not only in project cost, but also in the development of the principal contractor's construction phase plan. This must promptly respond, through controls exercised by safe and suitable systems of work. Ideally such design change is minimal.

Design changes during the construction phase are usually initiated by variation orders or architect's instructions. Such communication is dependent on timing, clarity, co-ordination and ownership.

6.2 Design change

The management of design change is critical in the interfaces between the design team/ CDM co-ordinator and the CDM co-ordinator/principal contractor. Such change acknowledges the sensitivity of management control of the construction process within the confines of the construction programme and puts the CDM co-ordinator in the role of ensuring that a suitable response is made to later design amendments on site.

These amendments to design will also need to have been subjected to the ongoing design risk management process and supporting documentation must provide the information needed.

Similar emphasis needs to be placed on the co-operation/co-ordination link between temporary and permanent design and on the need for good lines of communication between all the relevant parties and in particular with the overarching need for the client to discharge their duties under Regulation 9 ... 'the ongoing effectiveness of management arrangements.'

Section 7
DOCUMENTATION

The effective discharge of duty holder roles under the CDM Regulations 2007 ensures that all parties are contributors and communicators of health and safety information within the integrated team. It is important that the level of communication is appropriate so that key items of information are not lost and the charge of excessive bureaucracy that plagued earlier experiences under the CDM Regulations 1994 is duly countered.

The designer is entitled to base his level of communication on the premise that contractors are competent and hence there is no need to provide a focus on issues which a competent contractor would ordinarily expect to deal with. Concise, focused documentation is a vital tool in this communication linkage.

As highlighted in paragraph 131 of the Approved Code of Practice (ACoP), such information:

'should be project specific and concentrate on significant risks which may not be obvious to those who use design.'

Paragraph 132 further informs readers that:

'Designers also need to provide information about aspects of the design that could create significant risks during future construction work or maintenance. If in doubt about the level of information needed, the best way is to ask those who will use it.'

It is therefore prudent to consider the interpretation of the term 'significant' within a culture that is still on the cusp of change and alongside the interventionist strategies and implementation policies pursued by the HSE. The author has always found it more pragmatic for the design process to respond to significant and *principal* health and safety issues in the delivery of the compliant design response than to focus solely on significant issues. This offers the design team the ability to convey information about those significant categories of fatalities, including falls from height, contact with moving vehicles and falling objects. Additionally the occupational ill-health areas of manual handling, noise-induced hearing loss, hand–arm vibration, sensitory dermatitis and respiratory concern can be targeted irrespective of the competence of supply chain members. This extension of interpretation acknowledges the development of the process within the cusp of change that is part of the cultural movement within the construction industry.

The process of documentation is not only related to the transfer of information but also to the need for documentary evidence and transparency in respect of the design process itself. It must be based on the two perspectives of appropriateness and minimisation of paperwork, avoiding the bureaucratic nightmare associated with self-generating paperwork systems that are not read and which burden the recipient with information overload.

As stated in paragraph 4 of the AcoP:

'Any paperwork produced should help with communication and risk management. Paperwork which adds little to the management of risk is a waste of effort, and can be a dangerous distraction from the real business of risk reduction and management.'

This must accord with the HSE's message of:

'Managing risk and not paperwork'

The design process and its support systems are extensive and sophisticated and corresponding documentary evidence must range from the development of the client's brief to inputs for the health and safety file. There exists a portfolio of recording options, both in hard copy and software package format, but where deficiencies exist they must be quickly addressed. For the established design practice it is often better to modify existing systems than to implement wholesale change.

Documentary evidence and the design audit trail should merge with existing management systems and could be demonstrated in a variety of forms, including:

- agenda items
- design office manuals
- design philosophy statements
- design reports
- design risk assessments
- minutes of meetings, e.g.
  - brainstorming sessions
  - co-ordination meetings
  - design review meetings
  - scoping of works
- modelling
  - computer
  - three-dimensional
  - four-dimensional
- notes on drawings
- progress photographs
- project risk registers
- sequence drawings
  - three-dimensional
  - four–dimensional
- sequential erection statements
- tabulated forms

etc.

The management process is holistic and cannot be limited to any single form or tabulation. Agenda items, minutes and photographs are all crucial in providing documentary evidence of the effectiveness of design risk management. It should be remembered that the annotated note on the sketch or drawing can still provide the most effective and potent means of communication for the designer. The demonstration of compliance is facilitated greatly by the co-ordinating role of the lead designer.

As stated in the ACoP (paragraphs 144 and 145):

> *'Designers are not legally required to keep records of the process through which they achieve a safe design, but it can be useful to record why certain decisions were made. Brief records of the points considered, the conclusions reached, and the basis for those conclusions, can be very helpful when designs are passed from one designer to another. This will reduce the likelihood of important decisions being reversed by those who may not fully understand the implications of doing so.*
>
> *'Too much paperwork is as bad as too little, because the useless hides the necessary. Large volumes of paperwork listing generic hazards and risks, most of which are well known to contractors and others who use the design are positively harmful, and suggest a lack of competence on the part of the designer.'*

Thus, there is no legal requirement for the completion of any form, record or tabulation but the prudent design team should always have documentary evidence of their design risk management process. This is not an 'all singing, all dancing' record, but simply a framework of process visibility. Management control systems must support practitioners in discharging legal duties without enslaving them to a bureaucratic response.

There is no unique model for recording the design risk management process and it will depend on the nature of the undertaking, working practices, and the type and extent of the hazards and risks. Fundamental to the process is the risk assessment, which has an established pedigree in industries such as insurance, finance, aerospace, military, oil and gas. The approach in these industries has been developed for the wider remit and inevitably has accommodated aspects of finance and profit.

The petrochemical and process-plant industries have long-established risk models with incorporated health and safety dimensions and subsequently both these sources have offered much guidance to the construction industry itself.

Integral to the development of risk management in the construction industry itself was the workplace risk assessment perspective developed by contractors in the discharge of their duties under Regulation 3 of the Management of Health and Safety at Work Regulations 1999 (formerly 1992). Compliance with these rules became a requirement for all employers from 1 January 1993.

Design teams should distinguish between design risk assessment and workplace risk assessment. Design risk assessment accounts for those health and safety issues that the design team can influence because of their position in the process. Workplace risk assessment, meanwhile, involves a response to residual issues that are legitimately passed down the supply chain by the designer and others. Logically, where issues have been designed out, others further down the supply chain have nothing to manage. Where these issues have been reduced there is less for them to manage.

HSE Contract Research Report No. 71/1995[1] contains the following sections on risk assessment, which are still relevant today:

'There are a variety of methods of risk assessment, ranging from the crudely qualitative to the (relatively) sophisticated quantitative. Any method chosen will, to some degree, be subjective and arbitrary but, nevertheless, can prove useful provided it is appropriate for its purpose and its limitations are recognised.'

Various methods of risk assessment are set out in Table 7.1.

[1]Contract Research Report No 71.1995, *Construction (Design and Management) Regulations 1994 – Brief for a Designer's Handbook*, Health and Safety Commisssion.

Table 7.1

RISK ASSESSMENT METHODS

Qualitative	Quantitative
Brainstorming Delphi techniques HAZCON* Matrix methods Tabulations	Consequence analysis Decision trees Failure modes and effects analysis Fault tree analysis Management oversight and risk tree analysis. Probability analysis Sensitivity analysis Task analysis
Nomograms HAZAN** HAZOP*** studies	

*HAZCON, Hazards in Construction
**HAZAN, Hazard Analysis
***HAZOP, Hazard and Operability Studies

Brainstorming: a group approach used to generate a large number of ideas from the aggregate contribution of all team members. It is meant to stimulate creativity and lateral thinking.
Delphi technique: a brainstorming approach to decision making without necessarily bringing the contributors together. It is useful to move things towards a conclusion where logistically there would be difficulties in getting the participants around the table.
HAZCON: a formal procedure for the early identification and assessment of safety, occupational health and environmental risk in construction to enable all reasonable practical steps to be taken to reduce or eliminate them.
Matrix methods: refer to pages 93 and 104.
Tabulations: these can be qualitative or quantitative and can incorporate matrix methods or listings to provide visibility and ranking to the identified hazards (refer to pages 96–102; 110 and 114).
Consequence analysis: an essential part of risk assessment which models the potential consequences from activity failure. It can incorporate both consequential and cause events with all logical relationships (Cause Consequence Diagrams).
Decision trees: decision support tools that use a tree-like predictive graph or model of decisions and their possible consequences.
Failure modes and effects analysis (FMEA): a systematic team-driven approach that identifies potential failure modes in a system, product or manufacturing/assembly operation caused by design or process deficiencies. It is a tool to prevent problems occurring. Whilst developed by the military it now has wide use in the aerospace and automotive industries. Its purpose is to eliminate or reduce failures ideally at the design/development stage.
Fault tree analysis (FTA): another technique for reliability and safety analysis. It is a top-down approach that is basically composed of logic diagrams that display the state of the system and is constructed using graphical design techniques. It originated in the Bell Telephone Laboratories and was then used by the military, nuclear and aerospace industries. Also used in accident investigation.
Management oversight and risk tree analysis (MORT): a safety analysis method and is top-down fault tree based. The tree gives an overview of the causes of the top event from management oversights and omissions or from assumed risks or both. Although associated with nuclear power and aerospace industries it is also used in accident investigations.
Probability analysis: a statistical approach to the occurrence of possible outcomes often accompanied by confidence limits associated with the probabilities selected.
Sensitivity analysis: a mathematical modelling tool that considers the effect of input factor variations on the conclusions drawn from the model. It is useful where there are many input variables and contributes to the robustness of the result. Frequently used in physics, chemistry, financial analysis, signal processing and neural networks.
Task analysis: a breakdown of activities undertaken to complete the task. An applied behaviour analysis technique which can lead to the drafting of procedural statements.

Nomograms: a graphical representation via a two-dimensional diagram which is fundamentally an analogue computation device, similar in principle to a slide-rule.
HAZAN: Hazard Analysis is the application of numerical methods to obtain an understanding of hazards and is an essential pre-requisite for a risk assessment process. Such an analysis provides a sound quantitative basis for decisions on mitigation measures.
HAZOP studies: HAZard and OPerability studies provide a more detailed consideration of the safety and operability implications of a well defined design or procedure. It is a team-based, structured method of identifying hazards, contributory causes and operability problems in plant and procedures and is a standard risk management technique associated with the nuclear, petrochemical and process plant industries. As suggested by its name, its focus is on operational issues and provides mitigation and contingency response management to failure situations.

Not all of the methods shown in Table 7.1 are appropriate for design purposes but mainstream design approach tends to rely on combinations of brainstorming, matrix methods and tabulations.

The familiar matrix method in its basic form can be represented by the following relationship:

$$\text{risk} = \text{severity} \times \text{likelihood}$$

where risk is the manifestation of a hazard and hazard is defined as anything with the potential to cause harm. Conventionally, as shown, risk is the product of severity and likelihood, where severity ranges from death to an incident level and likelihood is the frequency or probability of occurrence.

Severity and likelihood factors can be rated numerically by assigning scores in ranges from 1 to 3, 1 to 5 and even 1 to 10. Alternatively, a qualitative approach can be used, assigning low, medium or high factors.

The following examples of corresponding documentation provide a further insight to the process.

Table 7.2

EXAMPLES OF POTENTIAL HAZARDS FOR DESIGNERS TO CONSIDER

Hazardous activity	Examples of hazard	Example of designer's intervention to aid control of the risk
Work at height	Fall from a flat roof	Design in parapet or barrier Design in provision to ease installation of temporary handrails
	Fall through fragile roof/skylight assemblies	Don't specify fragile materials Identify existing fragile assemblies Position ventilation and extraction equipment to avoid going on roofs
	Fall from ladder	'Design out' the need for ladders during construction, cleaning and maintenance operations, e.g. • design stairways for use during construction • design hard standing to allow use of mobile access equipment • design windows to be cleaned from the inside • specify materials that don't need routine painting, or design in safe access for maintenance • consider prefabrication so that sub-assemblies can be erected at ground level and then safely lifted into place
Working in or close to excavations	Poor ground conditions, resulting in collapse, inundation, asphyxiation, etc. Contact with contaminants	Provide adequate information about ground conditions and position of services Limit depth of excavation
Working close to plant and vehicles	Struck/trapped by moving plant or vehicles	Position structures to allow: • safe access and egress onto public roads • the minimising of reversing • the segregation of pedestrians and vehicles
Working on electrical systems	Exposure to live contacts Contact with overhead or underground cables	Identify existing service positions Position structures to minimise risks from: • buried services • overhead cables Design services so isolation is possible
Work on altering or erecting structures	Collapse of the structure due to instability	Suggest a sequence of erection Design sacrificial bracing elements to aid erection Design structure for erection loads State design philosophy and assumptions for stability Provide limitations on lifting sling angles Inform client of his duty to provide information, e.g. a structural survey
Working in unergonomic/ strenuous conditions	Inappropriate and repetitive manual handling	Specify light blocks (<20 kg) Design rebar cages for lifting Specify couplers in place of long laps, to aid steel fixing
	Handling heavy loads, e.g. kerb stones	Adapt design for the use of mechanical aids Specify lighter alternatives
	Lifting in awkward posture, e.g. needing to twist and turn, particularly when repeated	Design for ease of access, e.g. avoiding need for awkward postures or twisting in plant room Consider space requirements for access, e.g. services in ceiling voids, fixing rebar
Working with hand-held tools	Hand arm vibration syndrome (HAVS)	Specifying surface finishes that don't require scabbling Avoid chasing Avoid hand tunnelling Design piles so that mechanical pile cropping is possible

Hazardous activity	Examples of hazard	Example of designer's intervention to aid control of the risk
Working with hazardous materials	Exposure to irritants, corrosives, asbestos, biochemicals, radiological agents, toxins, etc, e.g. contact with wet cement	Specify low chrome cement Design to use bulk supply pumped concrete, to reduce skin contact Provide enabling works to allow welfare facilities to be installed at the start of a project Inform client of his duty to provide information, e.g. an asbestos survey Specify adhesives that have non-volatile solvents, e.g. water-based adhesives
Working in noisy environments	Noise, resulting in hearing loss	Adapt the design to allow the use of less noisy solutions, e.g. hydraulic piling Consider the use of self-compacting concrete Specify crack-inducers, where appropriate, to avoid saw cutting Cast in brick ties instead of shot-firing
Working in confined spaces	Asphyxiation, noise, inundation, etc.	Examine whether the design can avoid a confined space Make provision in the design for prompt and easy rescue Make adequate provision for access Avoid on-site welding
Work on restricted sites, e.g. refurbishment	Handling of heavy and unwieldy components	Use alternative structural sections, e.g. multiple rolled steel angles for a single universal beam Specify spliced beams

Note: Useful as a generic example of design contribution. Also refer to the CIRIA Report 166[2] and '*Safety in Design*'[3]

[2] CIRIA Report 166, *CDM Regulations – Work Sector Guidance for Designers*, CIRIA (F

[3] *CDM Guidance for Designers*, Construction Industry Council (November 2003).

7.1 Red, amber and green lists[4]

This is a 'traffic light' representation, where the red list represents prohibited action, amber categorises acceptable action provided it is supported by coherent arguments and green offers acceptable procedures to be followed.

Red lists

- Pre-construction information-tender stage (formerly Pre-tender health and safety plan) not to be issued until detailed structural surveys, asbestos surveys, etc. completed.
- Scabbling of concrete ('stop ends', etc).
- Demolition by hand-held breakers of the top sections of concrete piles (pile cropping techniques are available).
- The specification of fragile rooflights and roofing assemblies.
- Processes giving rise to large quantities of dust (dry cutting, blasting, etc.).
- On-site spraying of harmful particulates.
- The specification of structural steelwork that is not purposely designed to accommodate safety nets.
- Designing roof-mounted services requiring access (for maintenance, etc.), without provision for safe access (e.g. barriers).

Amber lists

- Internal manholes in circulation areas.
- External manholes in heavily used vehicle access zones.
- The specification of 'lip' details (i.e. trip hazards) at the tops of pre-cast concrete staircases.
- The specification of shallow steps (i.e. risers) in external paved areas.
- The specification of heavy building blocks, i.e. those weighing >20 kg.
- The specification of large and heavy glass panels.
- The chasing out of concrete/brick/blockwork walls or floors for the installation of services.
- The specification of heavy lintels (the use of slim metal or concrete lintels being preferred).
- The specification of solvent-based paints and thinners or isocyanates, particularly for use in confined areas.
- Specification of curtain wall or panel systems without provision for the tying of scaffolds.
- Specification of blockwork walls >3.5 m high and retarded mortar mixes.

Green lists

- Adequate access for construction vehicles to minimise reversing requirements (e.g. one-way systems and turning radii).

[4]With acknowledgement and thanks to Chris Fitt of Managing CDM Ltd and BAA Terminal 5, who provided several of these examples.

- Provision of adequate access and headroom for maintenance in plant rooms and adequate access provision for replacing heavy components.
- Thoughtful location of mechanical and electrical equipment, light fittings, security devices and so on, to facilitate access and away from crowded areas.
- The specification of concrete products with pre-cast fixings to avoid drilling.
- Specify half-board sizes for plasterboard sheets to make handling easier.
- Early installation of permanent means of access and prefabricated staircases with hand rails.
- The provision of edge protection at permanent works where there is a foreseeable risk of falls after handover.
- Practical and safe methods of window cleaning (e.g. from the inside).
- Appointment of a Temporary Work Co-ordinator (BS 5975).
- Off-site timber treatment if PPA- and CCA-based preservatives are used (boron or copper salts can be used for cut ends on site).

Note: This listing would be useful as a generic statement contributing to the modus operandi within a design office manual.

Table 7.3
DESIGN RISK ASSESSMENT

Hazardous activity	Residual hazard	Information provided for hazard control (illustrative)
Erecting floor steelwork	Falling from height off unguarded edges	60.3 × 5 CHSs provided at 3 m c/c on perimeter beams for insertion of 48.3 mm of edge protection posts Safety nets may be used (but see below). Note: Storey height only 3.7 m Staircase in A/B, 4/5 designed as freestanding between floors and may be used for access Holes in beam and column flanges provided as anchorages for lanyards (but see above re clearances)
	Premature collapse of structural component	Erection loads must not exceed 1.5 kN/m^2 anywhere Safety nets may not be attached to beams with lateral section properties $<305 \times 102 \times 33$ Ubs All perimeter trusses and vertical bracing to be in place before work starts
	Handling of heavy components Cranes	Perimeter trusses max weight = 1000 kg When lifting, angle between the slings should not exceed 30 degrees to avoid buckling of section Other main floor steel sections are 54 kg/m, with a maximum piece weight of 351 kg (sling angle: any permitted) Space (for cranes) clear of services at building along north and south perimeter. Assumed position gives maximum lifting radius = 19 m and maximum lifting height = 20 m. (Note: Overhead power lines to east side of building! See Drg. No. ••) Ground conditions: see SI Report, Ref. •• (page ••) CoG of beams specified to be marked by fabricator (will change if temporary edge protection attached before lifting) Shelf angles provided on main columns only All bolts 20 mm diameter Deliveries: See sheet x of y
	Electric shock overhead power	Note: Overhead power lines 10 kV at SE corner of site, crossing site at approx. 45–50 m in from site south boundary (See Drg. No. ••)

Observation: Project-specific with much detailed and useful information, indicative of effective dialogue between designer and contractor.

Figure 7.1

EXAMPLE OF A DESIGN RISK ASSESSMENT PROFORMA

ject: Various			Completed by: SDS							
tivity ement	Significant potential hazards	Population at risk	Risk classification			Design action to be for mitigation	Design action ownership		Reduced risk factors	Future action
			Likelihood	Severity	Risk rating		By	Date completed		
Sub-structure Piling	Industrial legacy of site. Contaminated material arising	Operatives	3	3	9	1. Geotechnical survey. 2. Driven pile	SDS	April 2001	1 × 3 = 3	PC* information H&S file**
External enclosure	Manual handling (component wts estimated as 50–60 kg min.)	Operatives	3	2	6	Mechanised handling Automation	SDS	April 2001	2 × 2 = 4	Method statement Specification PC information
Traffic management	Reversing vehicles	Operatives	Alternatively, consider simplification using high; medium or low:			One-way system, no reversing	SDS	April 2001	1 × 2 = 2	Signage CP Plan***
Superstructure	Instability (temporary) Wind limitations during erection	Operatives	High			Sequential erection sequence to be communicated. 20 mph wind limit.	SDS	April 2001	Medium	Method statement. PC information H&S file
Superstructure (plant units)	Work at height	Operatives	Medium			Prefabrication Ground floor assembly Modularisation	SDS	April 2001	Low	Specification H&S file
Roof structure	Fragility	Operatives Public Maintenance	High			Non-fragile roof structure using structural glass Methodology	SDS	April 2001	Low	Design details and specification. Maintenance methodology

Severity:1 to 3	Likelihood: 1 to 3	Risk rating: product of likelihood and severity.
6 and 9: Red zone	3 and 4: Amber zone	1 and 2: Green zone

Alternatively		
High	Medium	Low

*PC, principal contractor.
**H&S, health and safety.
***CP Plan, construction phase plan.

Figure 7.1 Example of a design risk assessment proforma.

Both of the above formats and variations on them can be found throughout the industry, but the numerical matrix in whatever range (3×3, 5×5, 6×6, 10×10) has generally given way in recent years to the simpler system based on high, medium or low rankings.

The designer should appreciate that there is no numerical rigour in the use of numbers, since the statistical scale of probability (likelihood) is actually represented by a range of between 0 and 1, where 1 represents 100% certainty. Thus the scales used should at best be seen as a quasi-numerical approach, useful for ranking but offering no advantage over the simpler and broader classification based on high, medium and low gradings.

Indeed there is an argument for the designer to leave out any form of grading altogether since the grading comes after the hazard has been identified and recorded in the hazard column. In other words the decision has been made about the significance of the hazard in recording it on the form and this identification must be followed by a design contribution based on the general principles of prevention.

The 'five steps to risk assessment'[5] approach advocated by the HSE for the workplace is as follows:

Step 1 Identification of the hazard
Step 2 Decide who might be harmed and how
Step 3 Evaluation of the risks
Step 4 Record your findings
Step 5 Review and revise.

Similarities can be seen in the format and contents of the above form, but it is important to note that the *design* risk assessment, whilst similar in objective to the *workplace* risk assessment, is not identical and therefore documentary evidence must reflect the differences, whilst still providing a suitable record.

The better forms, which support the design risk management process, include the following characteristics:

- Identification of the hazard.
- Establishment of ownership.
- Offer mitigation (based on the general principals of prevention).
- Provide evidence of contribution.
- Establish further communication.
- Identify the ensuing documentation trail by reference to method statements, pre-construction information, construction phase plans and health and safety files.

Whilst the design risk management process must identify all foreseeable hazards relevant to all the stages of the project, these will not all have the potential to become significant and principal. It is only the significant and principal hazards that need to be communicated to clients, designers, contractors and CDM co-ordinators. Records such as the ones above need to provide due visibility in these areas.

[5] *5 Steps to Risk Assessment*, INDG 163(rev 1), HSE 4/02.

Figure 7.2

ANNOTATED NOTES (HEALTH AND SAFETY) ON DRAWING

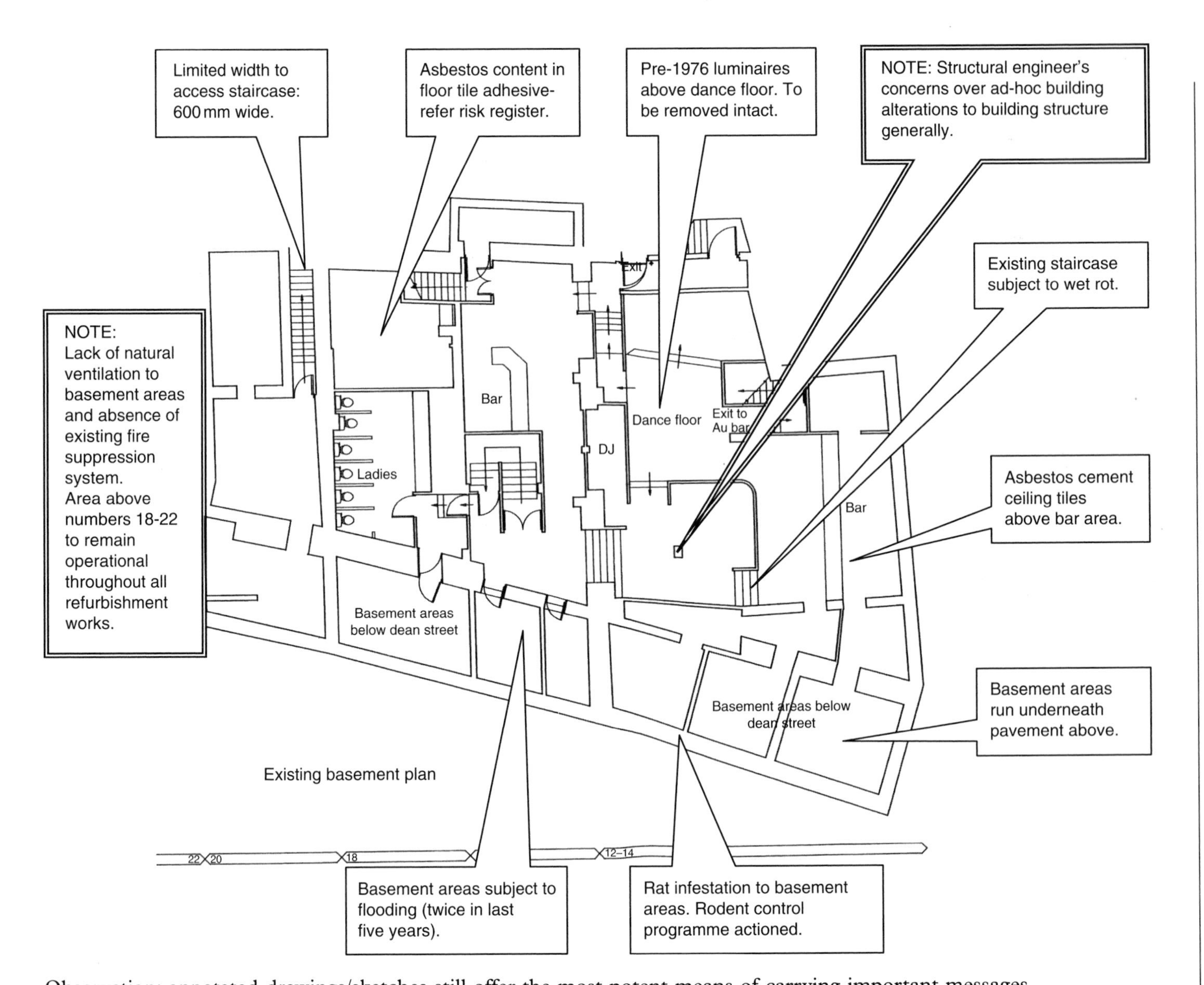

Observation: annotated drawings/sketches still offer the most potent means of carrying important messages.

Figure 7.2 Annotated notes (health and safety) on drawing.

Figure 7.3

HAZARD MANAGEMENT REGISTER AND DESIGN RISK ASSESSMENT

SCOTT+ BROWNRIGG

Project:			Job No: 13348	Workstage: A-B	Date: 24/01/07	Rev. E/1
Risk ref No. & Date opened	Element or activity & hazard or potential to cause harm	Persons at risk, likelihood of harm arising, and the consequences	Design team risk reduction proposals required to mitigate the risk (with options/alternatives recorded)	Responsibilities & actions: Action owner	Required date of action	Risk closed or residual
1.0	**Sitewide issues**	Injuries to contractors and operatives unaware of hidden features	Issues possibly affecting the site layout and infrastructural decisions on existing site. List of all surveys to be compiled and entered into H & S file	All design team members to advise	Prior to tender	**Open**
1.1	**Underground services-** Locations to be established in and around site. 2 No 36 inch gas mains, electrical and water mains, sewers, fibre optics, all to be located on survey. Existing services to QVC -not all provided in survey	Operatives-injuries during excavations existing users- gas explosions, electrical and flooding risks. Financial risks	Coordinated survey information required of entire site and surrounding roads, railways, highlighting risks to construction and needs for services diversions, transfer bridging structures, foundation relocation, or building footprint variations. Temporary construction access road could also affect services, as will car park depth and location. Loss of fibre optics main equates to £1mill. per day HL have consulted QVC & Statutory Authorities and awaiting feedback	AL, SB, HL	Prior to planning application design freeze	**Closed**
			Subscan survey completed revealing services locations and other large obstructions below ground eg. Underground diesel tank and large concrete bases to rear service road areas. Building and roads footprints to be overlaid for view of clashes	HL, SB	Detailed design analysis required at next stage prior to tender	**Open**
1.2	**Construction site-establishment-** Welfare, offices, parking, materials storage, crane locations, site roadways, cycleways	Operatives, existing users, pedestrians, other road users. Accidents and injuries due to large vehicular manoeuvres during and after construction	**Construction vehicular access and egress locations to be agreed with highways.** Site materials storage and on site vehicular distribution routes to be determined if at variance with eventual site infrastructure. Site compound location TBA temporary services may be required to welfare and office areas. Infrastructure and Phasing have impact. Possibility of influence on building and roads footprint	ECH, SB, HL, WSA	Prior to planning application design freeze	**Open**

Client		Architect	Scott brownrigg - **SB**	Structures		Services		
Proj. Man		Cost con		Landscape		P. Contractor		Highways		

Observation: A qualitative approach, with ownership, action dates and status of the issues all identified. No categorisation attempted, but focus throughout on issues that the design team can influence.

Figure 7.3 Hazard management register and design risk assessment. Reproduced with permission of Paul Bussey, Scott Brownrigg Ltd.

7.2 Project (health and safety) risk register

Whilst the previous forms and tabulations offer examples of industry practice, it is the author's contention that the concept of a project risk register offers numerous advantages. Project risk registers are a project management technique that have historically focused on issues that can jeopardise the project (usually financial). They are an accepted approach, and can be utilised for the management of health and safety issues and enable information to be presented in a collective sense which facilitates the promotion of the integrated team.

The project risk register (health and safety) is a collection of the significant and principal health and safety issues, collated from the project team members' inputs into one central document, with ownership, resolution times and outcomes all recorded. Its use enables the team to focus on all of the actions needed and, by review and suitable document management, on the developing strategies and methodologies required.

A project risk register readily reflects the iterative nature of project development, with treatment of issues remaining dynamic until delivery by the relevant party. The first contributions onto a risk register can usually be made by the client from source information, e.g. an existing health and safety file, with further contributions by the design teams, CDM co-ordinator, contractors and principal contractor compatible with project development. As its name suggests, it is a collection of relevant project issues, not just design issues, and so it represents a more holistic picture of the project and its development.

Ideally, the project risk register is owned by the project team and remains a dynamic document based on proactive input. In that sense its ownership sits with the project manager but this could be co-ordinated at suitable times by the lead designer, CDM co-ordinator and principal contractor. It embraces the project management philosophy and offers a more sophisticated approach than other records in accompanying the project rather than the discipline. It invites contributions from all duty holders, and health and safety management issues remain 'open' until closed out with concluding strategies, methodologies and duty holder responses.

Management control can be maintained by intranet access, with delivery of issues shaded for future monitoring and assessment. As a transparent document it also allows all duty holders to fully appreciate the health and safety parameters on which the project has developed.

As with all approaches it relies on the ability of the team to identify the issues and for effectiveness is dependent on a proactive approach and commitment by the team.

Table 7.4 illustrates the main features of a project risk register.

Table 7.4

PROJECT RISK REGISTER (HEALTH AND SAFETY)

Project risk register		Sheet	Date	Revision	Author	Status
		1 of 5	April 2009	0	As initialed	
Phase/activity	**Description**	**Hazard and/or risk description**	**Mitigation/controls**	**Ownership**	**Future action**	
Clearance	Basement areas	• Sharps and needles	• Advance sweep	• Client • Principal contractor	• Pre-start inspection. • Principal contractor information	Completed
Preparatory works	• Floor tile adhesive. • Versatemp panels • Ceiling tiles above bar area	• Asbestos	• AMP* • Remove • Registered contractor	• Principal contractor	• Principal contractor information • CP plan • H&S file	Methodology awaited
Structure	Roof-lights	• Falls from height (fragility).	• Designed out	• Designer	• Drawings • CP plan	Completed
Structure	Ad hoc repairs to walls between nos. 16 and 18.	• Collapse (str. integrity)	• Structural inspection • Sequential alterations	• Designer • Principal contractor	• Str. report • Principal contractor information • CP plan	Methodology awaited
	Steel columns and beams	• Fire/hot work	• Bolted connections. • Full height stanchions	• Designer • Principal contractor	• Specification • CP plan	Awaited
	External staircase	• Work at height • Falls from height • Manual handling	• Prefabrication • Bolted connection • Factory treatment	• Designer • Principal contractor	• Specification • Principal contractor information • CP plan	Methodology awaited
	External staircase/ continued	• Weight • Access restrictions	• Weight (1.20 tonnes) • Two no. pre-fab. sections	• Designer • Principal contractor	• Drawing details • Specification	Methodology awaited from principal contractor
	Staircase	• Collapse (dry rot infestation)	• Replace • Phasing of works • Demarcation	• Designer • Principal contractor	• Specification • Principal contractor information • CP plan	Development of CP plan
Service provision	All service runs	• Dust (respiratory disease)	• Surface mounted • Dry lining	• Principal contractor	• Principal contractor information • CP plan	Development of CP plan
Maintenance	Roof work generally Glass cleaning	• Falls from height	• Parapet rails • Self-cleaning glass	• Designer • Principal contractor • FM team	• Principal contractor information • CP plan • H&S file	Strategy being developed
Construction phase throughout	Operational interface with hotel	• Trespass/security	Liaison with: • hotel manager • principal contractor	• Principal contractor	• Daily contact required.	Ongoing

Contents should be read in conjunction with associated design philosophy statements.
*AMP, asbestos management plan.

7.3 Design philosophy statements

Other elements contribute to the corresponding audit trail.
Design statements are useful in providing an insight into the parameters of design and the premises on which design has progressed through its various stages. Such statements can then be considered for inclusion in pre-construction information and/or the health and safety file, for example:

- Strategic project management decision taken to re-cycle all material arising from the demolition phase and use it for piling platforms and roadways, thus minimising traffic movement off site.
- Early decision taken to adequately prepare lay-down areas for the receipt of all steel-work at north east area of the site.
- Early discussions with contractors enabled work at height to be done from mobile elevating work platforms (MEWPs) and telehandlers, limiting use of scaffolding.
- Footprint of building was minimised because of contamination – plant rooms located on roof, drop manholes used, common service routes provided.
- Substructure areas not tanked have been designed as water-excluding elements with control of crack widths.
- Columns have been designed as vertical unrestrained cantilevers, ignoring any restraint from the diaphragm action of the floors.
- Masonry panels subject to crowd loading are of reinforced blockwork construction.
- Full-height shear walls are designed as braced cantilevers to transmit loading down to foundation levels.
- Composite and in-situ concrete floors are designed to act as diaphragms to transmit wind and, more significantly, the horizontal loads resulting from the large cantilever of the middle tier stand back to the shear walls.
- Floors have been designed to support the collapse loading from the floor/terrace immediately above in the event of accidental damage. The steel and concrete frames ensure a limited area of collapse.
- The primary trusses directly support both the side tertiary trusses and the retractable roof trusses, and also provide a major transfer node for the load from the secondary trusses back to the mast structure.
- Integrity of the roof has been ensured following the loss of support from any one mast.
- Design throughout has attempted to optimise the use of prefabrication, ground-floor assembly and factory treatments.
- Enhanced specification to all protective systems to extend the period between re-applications.
- All concrete specified with low chromium content (to minimise risk of sensory dermatitis).
- Hot work minimised by friction grip bolts.
- Continuous flight augured piles used to mitigate noise limitation levels.
- Displacement piles used to mitigate handling of contaminated ground at depth.
- Continuous Flight Augured (CFA) piles used to mitigate effects of contamination at depth and proximity of water bearing aquifer.
- Passive ventilation utilised as mitigation measures in removal of radon (and methane) gas.

- Common service runs adopted to minimise methane gas migration route.
- All services to be surface mounted but masked behind stud partitioning.
- All glazed windows internally beaded to facilitate future glazing replacement.
- Window modules to be pre-glazed prior to installation to reduce exposure to work at height.
- Self-cleaning glass used in lift structure to reduce frequency of manual cleaning.
- All medium density fibreboard (MDF) materials to be factory drilled and machined prior to site delivery, to reduce exposure to carcinogens.
- Programming of works has accounted for the Christmas shopping period/January sales.
- All existing fragile roof-lights have been designed out.
- Light emitting diode (LED) lighting used where possible to extend longevity of components.

Philosophy statements are often developed in the early stages of the project. They should be passed promptly to the lead designer and/or CDM co-ordinator, in order for him to meet his duties in communication, co-ordination and management of documentation.

These statements provide a useful insight into the way that the design has developed. The more effective decisions arise from the project team's strategic response, facilitated by early dialogue between project team members and stakeholders generally.

Design review meetings

These are a fundamental part of the project and offer creative exercises in the iterative development from concept to detailed design, as well as providing a forum for the exchange of relevant health and safety information.

Paragraph 138 of the ACoP notes that:

'Regular reviews of the design involving all members of the design team are particularly important in making sure that proper consideration is given to buildability, usability and maintainability. When considering buildability, meetings should include the contractor so that difficulties associated with construction can be discussed and solutions agreed before the work begins. When discussing usability and maintainability, involving the client or those who will be responsible for operating the building or structure will mean that proper consideration can be given to the health and safety of those who will maintain and use the structure once it has been completed. Doing this during the design stage will result in significant cost savings for the client, as rectifying mistakes after the structure has been built is always expensive.'

In keeping with good project management principles, it is advantageous to extend invitations at the appropriate design review stage to all stakeholders who can contribute or who might be affected by design development. The outcomes of such meetings should be recorded, including details of ownership of delivery and action dates.

Design reports

At critical stages determined by the health and safety milestones/gateways that have been established for the project, the design process can deliver much of what has been described in this section via interim and final reports. As with any other aspect of the project management process, key areas can be addressed via such a document, allowing time for perusal and providing a higher profile for essential concerns.

Section 8
INFORMATION FLOW

The infrastructure support for health and safety design risk management delivery is heavily dependent on communication, with failure to effectively communicate acknowledged as a consistent theme in the inability of teams to deliver project success. This failure can equally affect the passage of information between and within teams and is a factor requiring vigilance by the design manager.

The philosophy behind the CDM Regulations 2007 requires all duty holders to communicate on relevant matters of health and safety management and to contribute to the essential documentation that captures related issues. Thus the Regulations explicitly require an effective communication infrastructure between all duty holders, with information flow as essential for CDM management, as it is for project management.

This suggests that for important exchanges, information must be transmitted by more than one medium in order to get key messages across to the recipient. Reliance on tabulations, notes on drawings, design philosophy statements, design risk assessments and project risk registers individually may not always facilitate the achievement of corresponding objectives. However, the statutory duty to promptly provide health and safety information to others remains an obligation for all, and since the communication system within any project is a fundamental of management, there is a particular relevance for the client who has an ongoing duty to ensure the effectiveness of management arrangements throughout the project period (Regulation 9(2)) (refer to pages 81 and 82).

The theme of communication runs heavily throughout the CDM Regulations 2007, with duty holder responsibilities, interface proactivity, integration and associated documentation all established to expedite it. Nonetheless, these remain challenging areas, with evidence of non-compliance reminding us all of the need for improvement. Table 8.1 identifies and articulates the communication links that affect the design process together with the corresponding regulation.

Table 8.1
COMMUNICATION LINKS

Regulation	Description	Duty holder	Comment
5	Co-operation	All	Foundation of all communication
6	Co-ordination	All	Concept of the 'lead designer' facilitates information flow
10	Pre-construction information	Client	To be provided to '*every person designing the structure and every contractor who has been or may be appointed by the client*'
11(6)	Sufficient information	Designer	'*... take all reasonable steps to provide with his design sufficient information about aspects of the design of the structure or its construction or maintenance ...*'
18(2)	Sufficient information	Designer	'*... take all reasonable steps to provide with his design sufficient information about aspects of the design of the structure or its construction or maintenance as will adequately assist the CDM co-ordinator ...*'
20(2)(b)(ii)	Pre-construction information	CDM co-ordinator	'*... promptly provide in a convenient form to- every person designing the structure, and every contractor who has been or may be appointed by the client*'
20(2)(d)	Co-operation	CDM co-ordinator	'*take all reasonable steps to ensure co-operation between designers and the principal contractor during the construction phase in relation to any design or change to a design*'
22(1)(b)	Design change	Principal contractor	'*... liaise with the CDM co-ordinator in performing his duties during the construction phase in relation to any design or change to a design ...*'

These communication links are further monitored as part of the client duties under Regulation 9 in order to ensure the effectiveness of management arrangements throughout the project. They are further facilitated by the advice and assistance provided by the CDM co-ordinator (Regulation 20 (1)(b)).

Concerns have always been expressed at the tendency for a bureaucratic response to information provision. Such criticism is often a comment on managerial abdication rather than a fault of the Regulations themselves. The basis of information flow is that designers have a right to expect that they are dealing with competent duty holders and hence there is no merit in providing mundane information of little relevance to the process of informing.

The provision of information focusing on issues over and above what the parties would normally expect to deal with greatly helps in filtering out irrelevance through the delivery of succinct information. Such information exchange must also be project specific, with the avoidance of generic approaches.

As noted in the Approved Code of Practice (ACoP), paragraph 112:

> *'Where significant risks remain when they have done what they can, designers should provide information with the design to ensure that the CDM Co-ordinator, other designers and contractors are aware of these risks and can take account of them.'*

However the focus on *significant* risks remains a challenge to all particularly within a culture that continues to evolve. As previously mentioned, the author has found it easier to focus on *significant and principal* issues since this can embrace the cultural difference and acknowledge the politics of intervention. It is prudent to give visibility to the current initiatives of the Health and Safety Executive (HSE) as well as other relevant matters, but there is no merit or necessity in producing paperwork for the sake of it.

Stuart Nattrass (former chief inspector of construction at HSE) reminded us all that the usefulness of paperwork is in inverse proportion to the amount. This message stills carries credibility and serves as a reminder that all duty holders must remove the unnecessary and the irrelevant. Thus, in pure communication terms:

Less is more.

Apart from the design team's internal systems the pertinent communication vehicles identified in the CDM Regulations are:

- pre-construction information
- construction phase plan
- health and safety file.

These are all integral within the communication framework and require a proactive design team response to provide such relevant information to other duty holders *promptly*. Designers' document management systems must be appropriate to the task and they must receive/exchange information in a proactive way, regardless of the efficiency or otherwise of the other duty holder arrangements.

In terms of accountability and transparency, information exchange stands as partial evidence of the discharge of duties, or conversely as evidence of failure.

8.1 Pre-construction information

Pre-construction information should be seen as a process akin to that of an information highway, gathering and collecting relevant information throughout the project, with specific information streaming off to all those designing the structure and to contractors who have or might potentially be appointed by the client. It therefore remains an iterative and dynamic process, which is initially managed by the client (Regulation 10) and then, on the notifiable project, by the CDM co-ordinator (Regulation 20(2)(b)).

This information flow (refer to Figure 8.1) emanates from the client and initially takes the form of source material that provides an input to the desk-top study during the early design process. Thus existing health and safety files, condition surveys, asbestos management plans and historical site information are provided as part of the early procurement process to form initial pre-construction information. Such information is provided on the basis of reasonable enquiries having been made by the client and his support team.

Thereafter, the information stream gets 'wider and deeper', with the design process contributing to the inward flow of additional information as a result of design development and the outward stream as relevant information goes off to other designers and contractors who are tendering to the client. The CDM co-ordinator, once appointed ('as soon as practicable after initial design'), then becomes responsible for managing this process facilitated by the role of the lead designer.

Paragraph 56 of the ACoP notes that:

'Where design work continues during the construction phase, the pre-construction information will need to be provided to designers before work starts on each new element of the design.'

Paragraph 58 further informs readers that:

'The pre-construction information provided should be sufficient to ensure that significant risks during the work can be anticipated and planned for. It should concentrate on those issues that designers and contractors could not reasonably be expected to anticipate or identify, and not on obvious hazards such as the likelihood that the project would involve work at height. Appendix 2 lists topics that should be considered when drawing up pre-construction information.'

As indicated in Figure 8.1, the provision of pre-construction information could extend well into the construction phase, particularly on design and build projects where providers of specialist work packages with a design input are appointed well after the construction phase has started. Pre-construction information should be viewed as information required before that item of work is undertaken and not before the construction phase itself starts.

Appendix 2 of the ACoP provides a template for pre-construction information, although it fails to promote the information process to which Regulation 10 relates and appears to be compromised by the baggage of the 1994 Regulations.

The designer is advised to note its introductory paragraph:

'Information should be included where the topic is relevant to the work proposed ... The level of detail in the information should be proportionate to the risks involved.'

Figure 8.1
INFORMATION FLOW

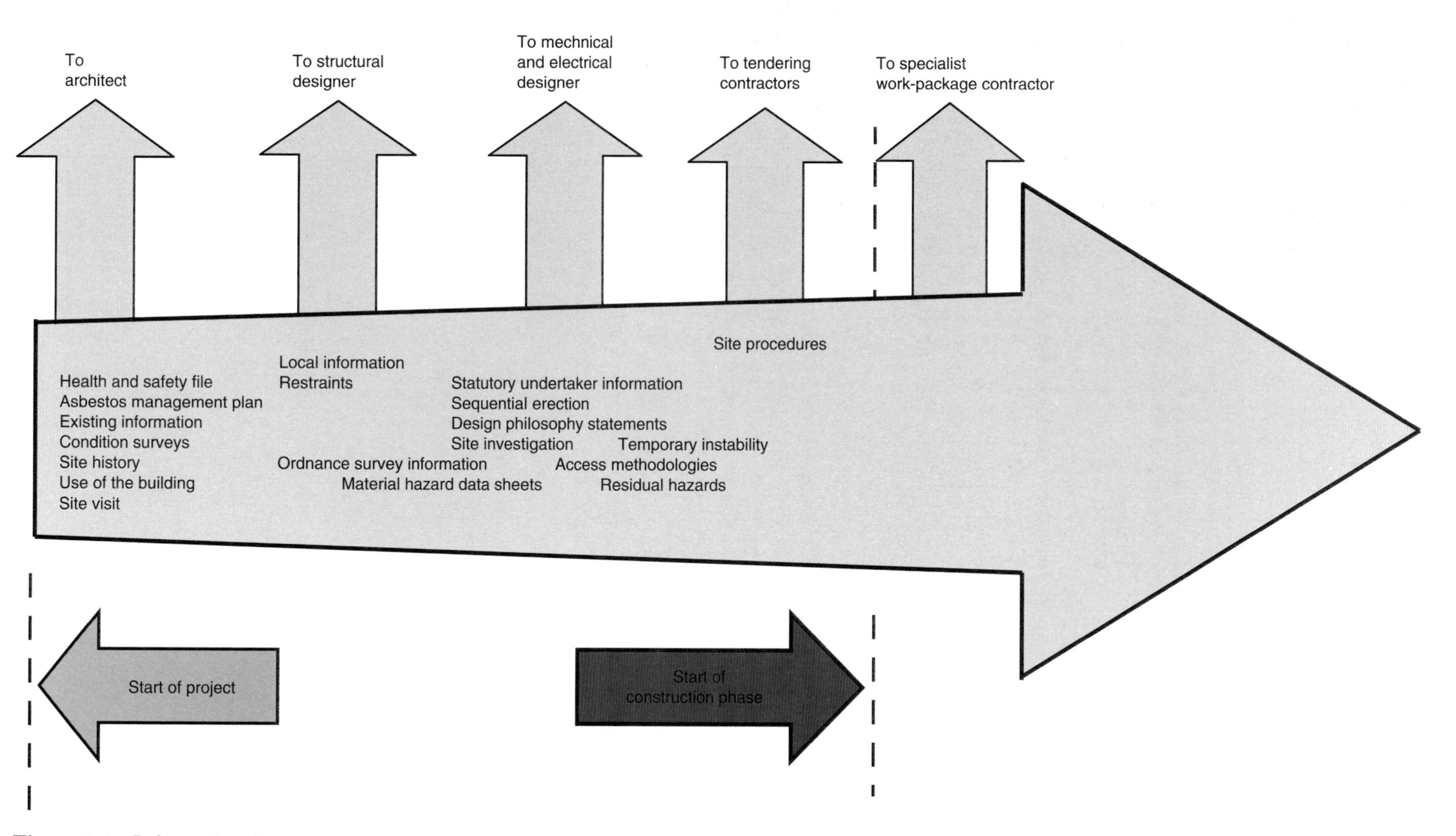

Figure 8.1 Information flow.

Early information under the heading of 'pre-construction information' must be given to those design teams appointed up-front in the process. The template is useful, but most clients would need guidance here and since this information is required prior to the appointment of the CDM co-ordinator, even on a notifiable project it is the initial design contact that must advise on the early collection and cataloguing of such information. The extent and limitations of information provided would then need to be verified by the CDM co-ordinator once appointed.

The design process contributes greatly to the information flow and much pre-construction information originates and is distributed during design and construction development. For the smaller project, existing paper-based systems will suffice, but on larger, more complex, multi-disciplinary projects the use of computerised management systems must be considered.

Designers should also note that their information systems must deal with non-notifiable projects and still get the right information to the right people at the right time. This may be challenging because of the absence of a CDM co-ordinator.

Figure 8.2 identifies those duty-holder interfaces and the corresponding Regulations that are critical for information exchange (See also Plate 2 in the colour plate section).

Refer to Figure 4.2 (page 54), where items on the left of the figure can be seen as inputs to the pre-construction information stream and those on the right as outputs.

Pre-construction information that goes out to tendering contractors demands contributions from the various design teams, who provide residual hazard information. These inputs focus on significant and principal issues, over and above what a competent contractor would expect to deal with. This is important, since it allows those tendering to define and resource the associated health and safety management controls. As such, these corresponding resources should then be included within the tender price submitted by the contractor.

In principle, this would be the case on both non-notifiable and notifiable projects, but there is usually more information to be relayed by design teams in the latter case.

Figure 8.2

DESIGN INTERFACE WITH OTHER DUTY HOLDERS

Designer duties		Interfaces with other duty holders				
Duty holder	Designers' duties	Client	Other designers	CDM co-ordinator	Principal contractor	Contractor
Duties in this row may establish interfaces with the duty holder and the designer	Regulations	9 10(1)		20(2);(b) 20(2)(c) 20(2)(d) 22(1)(b)*	13(2) 22(1)(a) 22(1)(b)*	13(2)
Designer	4	Competence	Competence		Competence	Competence
	5		Co-operation	Co-operation	Co-operation	Co-operation
	6		Co-ordination	Co-ordination	Co-ordination	Co-ordination
	7					
	11(1)	Client awareness				
	11(2)					
	11(3)					
	11(4)					
	11(5)					
	11(6)	Information	Information		Information	Information
	12		Design outside GB			
	18(1)	CDM co-ordinator appt.				
	18(2)			Information		
	25(2)**				Construction work	Construction work

Note: * Regulation 22(1)(b) requires liaison between the principal contractor and the CDM co-ordinator in respect of design or changes to a design

** Regulation 25(2) reminds, that any person other than a contractor who controls the way in which a contractor has to work could incur duties in respect of the management of the construction work (Regulations 26 to 44). In other words it is imperative that the designer does not impose construction methodology onto the contractor unless dictated by operational necessity

Figure 8.2 Design interface with other duty holders. See also Plate 2 in the colour plate section.

8.2 Construction phase plan

The construction phase plan is a key control document and is the responsibility of the principal contractor. It is associated with the notifiable project. Its development and adequacy is a prerequisite for the start of construction and must satisfy the client in this respect (Regulation 16). The client's duty to control the start of construction relies heavily on the advice and assistance of the CDM co-ordinator (Regulation 20(1)(a)), but nonetheless remains the client's to discharge.

The construction phase plan is an articulation of management controls to achieve *safe* and *suitable* systems of work, based on relevant aspects of construction management and the pre-construction information that went out to contractors tendering to the client. This pre-construction information forms the basis of principal contractor response, with solutions provided through the detail within his construction phase plan.

Regulation 20(1)(c) imposes a duty on the CDM co-ordinator to:

'liaise with the principal contractor regarding:

(ii) the information which the principal contractor needs to prepare the construction phase plan ...'

This subsequently rebounds back to the designer, who, by virtue of Regulation 11(6), is involved in that:

'The designer shall take all reasonable steps to provide with his design sufficient information about aspects of the design of the structure or its construction or maintenance as will adequately assist:

clients;
other designers; and
contractors

to comply with their duties under these regulations.'

and by Regulation 18(2) in that:

'The designer shall take all reasonable steps to provide with his design sufficient information about aspects of the design of the structure or its construction or maintenance as will adequately assist the CDM co-ordinator to comply with his duties under these regulations, including his duties in relation to the health and safety file.'

The designer, through his key contractual role, becomes an essential and critical link in this communication chain. The design team's communication and co-operation infrastructure must serve the needs of the project to ensure that the *right information gets to the right people at the right time.*

Regulation 11(6) also impacts on the interface between permanent and temporary design and it is important that the lead designer and CDM co-ordinator are satisfied that due lines of co-operation and co-ordination are implemented in respect of duties associated with Regulations 5 and 6. The over-arching duty remains on the client to ensure the ongoing effectiveness of management arrangements. Indeed the link between permanent and temporary design must be clear and well established for essential information to flow between designer and contractor at the critical stage of construction activity and in respect of design change.

The prevalence of design-and-build forms of contract ensures that permanent design is never complete at the start of the construction phase and hence the link between lead designer and CDM co-ordinator is even more vital. Even in conventional forms of admeasurement contract, despite expectations to the contrary, design is rarely complete and design flexibility is necessary in order to accommodate inevitable changes. Such change can create further challenges to site management because of the contractual need to accommodate change within the confines and constraints of a 'live' construction programme.

Design teams need to be vigilant to this link, which, under Regulation 11(6), further extends designer duties to the 'sharp' end of the construction process. Regulations 22(1)(b), 20(1)(c)(iii) and 20(2)(d) endorse this need, creating further duties for the principal contractor and the CDM co-ordinator respectively, namely that they must ensure that any such design changes receive a timely response from the principal contractor.

8.3 Health and safety file

This is a client document whose purpose is to:

> *'contain the information needed to allow future construction work, including cleaning, maintenance, alterations, refurbishment and demolition to be carried out safely.'*

There are two aspects of this document for the designer to consider, and these are outlined in the following sections.

Existing health and safety file

Where it exists, the health and safety file constitutes source information to be accessed as part of the designer's initial desk-top study and provides input to the pre-construction information process.

There will be occasions when such documentation doesn't exist or the information within is not current.[1] In these circumstances the designer will then have to advise the client on the need to undertake further surveys or investigations in order to supply the necessary data so that the design can be progressed. The cost factor here could be excessive.

Note: Design work on a non-notifiable project could lead to a modification/amendment to an existing health and safety file. The existence of the health and safety file needs to be established for all projects and its identification and currency are early design questions to be asked of the client.

[1] Regulation 17(3)(b): '*The client shall take all reasonable steps to ensure that after the construction phase the information in the health and safety file is revised as may be appropriate to incorporate any relevant new information.*'

New notifiable projects

The CDM co-ordinator's duty in respect of the health and safety file is to:
Regulation 20(2)(e) ...

'... prepare, where none exists, and otherwise review and update a record ('the health and safety file') containing information relating to the project which is likely to be needed during any subsequent construction work to ensure the health and safety of any person, including information provided in pursuance of regulations 17(1), 18(2) and 22(1)(j) ...'

Information from the design teams constitutes an essential input to the substance and purpose of any health and safety file. Early dialogue should be established to identify such inputs and later frustrations will be minimised if a clear understanding of contributions is determined. Whilst designers are aware of the requisite communication of residual hazards into this document there are still CDM co-ordinator subjective decisions to be made on the portfolio of drawings required. Early dialogue reduces misunderstandings.

The design team should note that this document is meant to be handed over at 'the end of the construction phase' and yet the enduring criticism made of all duty holders is their failure to deliver relevant information to the CDM co-ordinator for the completion and handover of the health and safety file at the appropriate time. The design manager's role is to pass over information as and when available. This demands proactivity on the part of all duty holders and there is much improvement required if health and safety files are to be delivered in a manner that is compatible with project handover to the client.

The design information in Table 8.2 should be considered as potential inputs to the health and safety file where relevant.

Table 8.2

DESIGN INFORMATION FOR THE HEALTH AND SAFETY FILE

Item	Comment
Calculations for the purposes of design	Loadings (unusual), hydrological, pressure, seismic, slope stability, structural, tidal flows(traffic)
Design philosophy statements	Useful appreciation of some of the rationale to the design process from a health and safety perspective
Information	Unusual characteristics that might influence future modifications, e.g. demolition and high density concrete Equipment/plant information critical for removal/dismantling
Maintenance regimes	Critical for developing testing frequencies for structural integrity, e.g. connection points for abseiling, barrier testing (sports grounds), support points for maintenance cradles, etc.
Material hazard data sheets	Residual hazard information to allow future management to make their own COSHH assessments in respect of articles/materials/substance
Methodologies	For consideration in respect of specialised systems such as: • access/maintenance equipment • replacement • window cleaning
Record drawings	'As-built' and/or 'as-laid' drawings to allow necessary future reference to be made
Residual hazard information	Essential insight to management factors associated with future work, etc.
Sequential erection statements	Essential for temporary stability during both erection and disassembly
Structural principles	Essential for future alterations/modifications and demolition appreciation
Workplace (Health, Safety and Welfare) Regulations 1992	
Features	Relevant
Materials used	Residual hazard information essential for replacement.
Maintenance methodology	Necessary for accessing and maintaining

Nonetheless it remains an indictment on the design process if such information cannot be provided at the appropriate time. Early discussions between the client, designer and CDM co-ordinator are recommended in order to determine the format and timing of information delivery.

It is not uncommon for the principal contractor on a design-and-build project to install his own design co-ordinator within the management structure of the design team. This enables better communication and establishes a direct link with design progress and delivery schedules. It also counters some of the concerns expressed by contractors in respect of novation.

The inheritance of a design team that has initially been appointed by the client can disrupt the continuity and integration of the project team. Good communication can overcome some of the related anxieties.

The humble bar chart (Figure 8.3) epitomises effective communication. It is simple, universally accepted, easily understood and capable of monitoring and review, all of which are characteristics of good communication and which have contributed to its longevity, provided its limitations are appreciated.

If design teams are intent on promoting integration and controlling the management of information then consideration could be given to the utilisation of the bar (GANTT) chart as a means of highlighting key dates and the linkage of process events. The necessary feed of design information into the trigger points of tender documents; pre-construction information (tender stage), and health and safety files can all be illustrated for the benefit of managerial control and action (see below). Subsequently, they can be highlighted on the corresponding network approaches to programming.

Figure 8.3

PLANNING/PROGRAMMING INTEGRATION

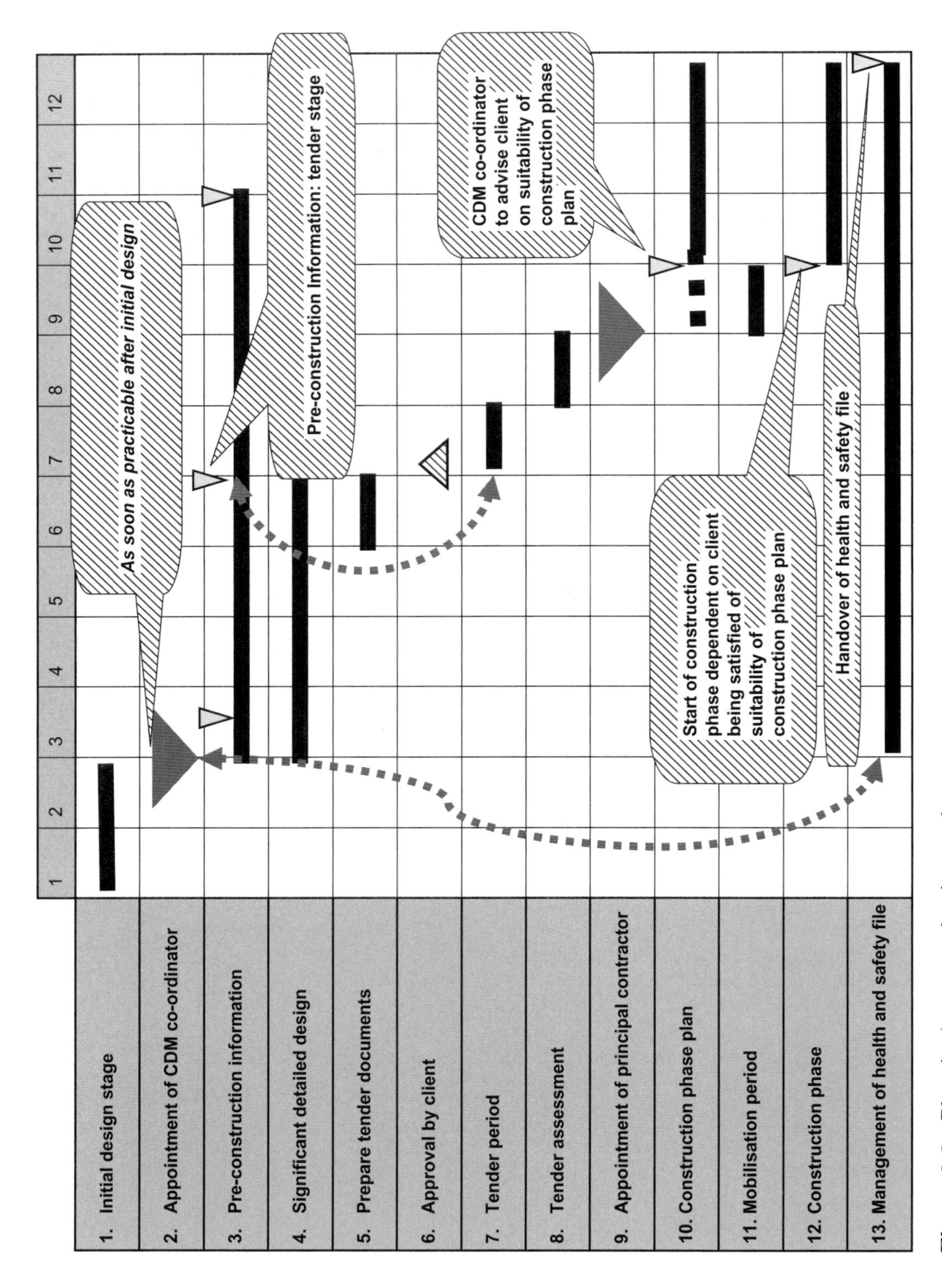

Figure 8.3 Planning/programming integration.

Appendix One
ROADMAP

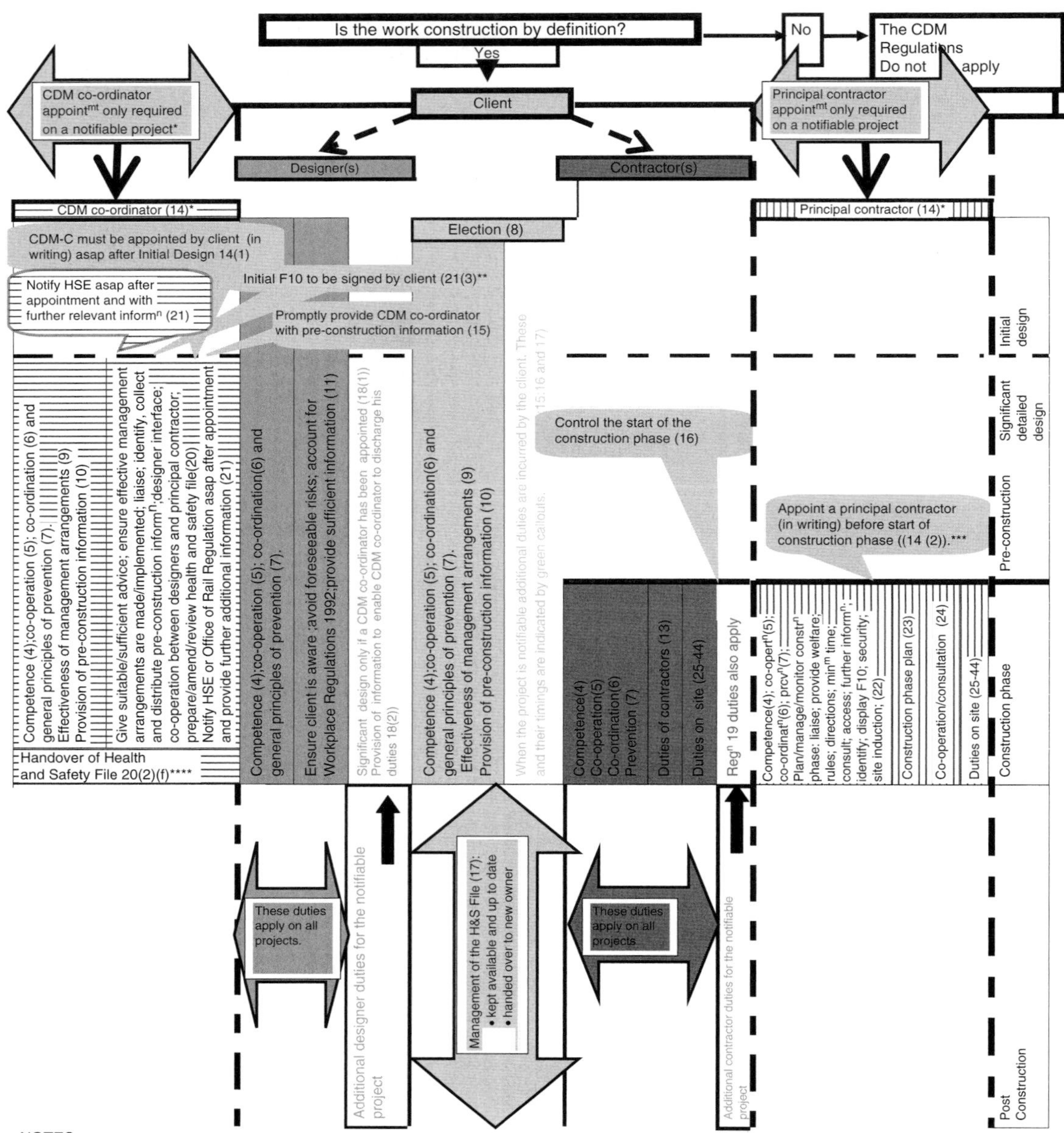

NOTES:

*Construction projects become notifiable if the construction phase exceeds 30 days or more than 500 person days. (refer Regulation 2(3)).

** Notification to the HSE/Office of Rail Regulation to be signed by client or someone on his behalf (Schedule 1)

*** It is recommended (regulation 14(2)) that the principal contractor is appointed as soon as practicable after the client knows enough about the project to be able to select a suitable person (i.e it depends on the procurement strategy)

**** Health and safety file to be handed over by CDM co-ordinator at end of the construction phase

Legend:

Client	Duties to be undertaken by clients on all projects
Client	Additional duties to be undertaken by client on notifiable projects
Control the start of the construction	Additional duties and timings to be undertaken by client
Designer	Duties to be undertaken by designers on all projects
Designer	Additional duties to be undertaken on notifiable projects
Contractor	Duties to be undertaken by contractors on all projects
Contractor	Additional duties to be undertaken on notifiable projects
CDM co-ordinator	Duties to be undertaken by CDM co-ordinator on notifiable projects
Principal contractor	Duties to be undertaken oby principal contractor on notifiable projects

S.D.Summerhayes/June 2009

For a colour version of this figure, see Plate 3 in the colour plate section.

Appendix Two

REFERENCES AND BIBLIOGRAPHY

A Guide to the Control of Substances Hazardous to Health in Design and Construction, CIRIA Report 125, CIRIA (1993).

Backs for the Future, Safe manual handling in construction, HSG 149, HSE (2000).

CDM 2007 – Construction Work Sector Guidance for Designers, 3rd edn, CIRIA Report 662.

CDM Guidance for Designers, Safety in Design. www.safetyindesign.org.

CDM Regulations – Work Sector Guidance for Designers, CIRIA Report 166, CIRIA (February 1997).

CDM Regulations – Construction Work Sector Guidance for Designers, CIRIA Report 662, Ove Arup, CIRIA (2007).

CDM Regulations Procedures Manual, Summerhayes SD, Blackwell (2008).

Control of Substances Hazardous to Health, 5th edn, Approved Code of Practice and Guidance L5, HSE Books (2005).

Design Risk Management, Smith NC, RIBA Publishing (2007).

Egan, Sir John, *Re-thinking Construction*. HMSO, London (1998).

Engineering Construction Risks, Thompson P and Perry J, Thomas Telford Services Ltd (1992).

Final Report of the Advisory Committee on Falsework, HSE, HMSO (1976).

Five Steps to Risk Assessment, INDG 163 (rev 1), HSE (4/02).

Health and Safety at Work etc Act 1974, reprinted with corrections 2006, The Stationary Office.

Latham, Sir Michael, *Constructing the Team*. Final Report of the Government Industry Review of Procurement and Contractual Arrangements in the UK Construction Industry. HMSO, London (1994).

Management of Health and Safety at Work Regulations 1999, Approved Code of Practice and Guidance L21, 2nd edn, HSE Books (2001).

Managing Health and Safety in Construction, Approved Code of Practice (L144), HSC, HSE Books (2007).

Manual Handling, Guidance on Regulations L23, HSE Books (reprinted 1998).

Mitchell B and Trebes B, *NEC Managing Reality Book 3: Managing the Contract*, Thomas Telford Books (2005).

New Civil Engineer, 23 September 2004.

Newsletter: Issue 38, Association for Project Safety (formerly the Association of Planning Supervisors) (2004).

Peer Review of Analysis of Specialist Group Reports on Causes of Construction Accidents, Research Report 218, HSE Books (2004).

Practice Notes (numerous), The Association for Project Safety.

Regina *v.* Paul Wurth SA, Court of Appeal (26 January 2000).

Report of the Summerland Fire Commission, Government Office, Isle of Man (May 1974).

Risk, Management and Procurement in Construction, Centre of Construction Law and Management Seventh Annual Conference (1994).

Safety by Design, An Engineer's Responsibility for Safety, Hazards Forum (1996).

The Abbeystead Explosion, Health and Safety Executive, HMSO (1985).

The Collapse of NATM Tunnels at Heathrow Airport, HSE Books (2000).

The Construction (Design and Management) Regulations 2007, Statutory Instruments 2007 No 320, HMSO.

The Control of Noise at Work Regulations 2005, 2nd edn, Guidance on Regulations (L108), HSE (2005).

The Long Term Costs of Owning and Using Buildings, Report of the Royal Academy of Engineering (1998).

The Report of the Court of Enquiry, signed by W. Yoland and W.H. Barlow (1880). *An Addendum to the Enquiry Report*, signed by H.C. Rothery (1880).

The Work at Height Regulations 2005, Statutory Instrument 2005 No. 735, TSO.

Walkway Collapse at Port Ramsgate. A Report on the Investigation into the Walkway Collapse at Port Ramsgate on 14 September 1994. HSE Books, HMSO (2000).

Workplace (Health, Safety and Welfare) Regulations 1992, Approved Code of Practice L24, HSE Books (1992).

Workplace Health, Safety and Welfare, Approved Code of Practice L24, 11th impression, HSE Books (1997).

Workplace – 'In-use' Guidance for Designers, CIRIA Report 663, CIRIA (2007).

Appendix Three
WEB PAGE DIRECTORY

No	Address	Description
1	www.aps.org.uk	The Association for Project Safety
2	www.bre.co.uk	Building Research Establishment
3	www.cibse.org	Chartered Institute of Building Services Engineers
4	www.thecc.org.uk	Construction Confederation
5	www.constructingexecellence.org.uk	Constructing Excellence
6	www.cic.org.uk	Construction Industry Council
7	www.citb.org.uk	Construction Industry Training Board
8	www.cpa.uk	Construction Plant Hire Association
9	www.dti.org.uk	Department of Trade and Industry
10	www.agency.osha.eu.int	European Agency for Safety and Health at Work
11	www.hse.gov.uk	Health and Safety Executive: United Kingdom
12	www.hsebooks.com	Health and Safety Executive Books
13	www.hsa.ie	Health and Safety Authority: Ireland
14	www.ice.org.uk	The Institution of Civil Engineers
15	www.imeche.org.uk	The Institution of Mechanical Engineers
16	www.istructe.org.uk	The Institution of Structural Engineers
17	www.nao.org.uk	National Audit Office
18	www.demolitionhyphen;nfdc.com	National Confederation of Demolition Contractors
19	www.nhbc.co.uk	National House Building Confederation
20	www.ogc.gov.uk	Office of Government Commerce
21	www.eustatistics.gov.uk	Office for National Statistics
22	www.pff.org.uk	Precast Flooring Federation
23	www.rias.org.uk	The Royal Incorporation of Architects in Scotland
24	www.architecture.com	Royal Institute of British Architects
25	www.rics.org	The Royal Institute of Chartered Surveyors
26	www.rospa.com	The Royal Society for the Prevention of Accidents
27	www.safetyindesign.org	Safety in Design Ltd
28	www.scoss.org.uk	Standing Committee on Structural Safety
29	www.strategicforum.org.uk	Strategic Forum
30	www.design4health.com	University of Loughborough
31	www.tso.co.uk	The Stationary Office
32	www.hse.gov.uk	Health and Safety Executive

Appendix Four

WORKPLACE (HEALTH, SAFETY AND WELFARE) REGULATIONS 1992

Note that these regulations aim to ensure that workplaces meet the health, safety and welfare needs of each member of the workforce, which may include people with disabilities.

No.	Regulation	Comment
1	Citation and commencement	**All such workplaces must now comply.**
2	Interpretation	**Applies principally to factories, shops, offices, schools, hospitals, hotels and places of entertainment.** **Also includes common parts of shared buildings, private roads and paths on industrial estates and business parks, and temporary work sites (but not construction sites).**
3	Application of these regulations	Applies to all workplaces except in or on a ship, activities relating to building operations or works of engineering construction within. All operational ships, boats, hovercraft, aircraft, trains and road vehicles are excluded. They do not apply to underground workplaces at mines, quarries or other mineral extraction sites, including those off-shore. Construction sites (including site offices) are excluded. Where construction work is in progress within a workplace, it can be treated as a construction site if it is fenced off and excluded. If not these regulations and the CDM Regulations 2007 will apply.
4	Requirements under these regulations	Every employer and those having control of a workplace shall ensure that every workplace is compliant with these regulations. Note: Employers have a general duty under Section 2 of the HSAW Act 1974 to ensure, so far as is reasonably practicable, the health, safety and welfare of their employees at work.
5	Maintenance of workplace, and of equipment, devices and systems	Such items shall be maintained (including cleaned as appropriate) in an efficient state, in efficient working order and in good repair. There is the requirement for a suitable system of maintenance, where appropriate, for certain equipment and devices and for ventilation systems. **Particular emphases on: emergency lighting, fencing, fixed equipment for window cleaning, anchorage points for safety harnesses, devices to limit the opening of windows, powered doors, escalators and moving walkways.**
6	Ventilation	Every workplace to be ventilated by a sufficient quantity of fresh or purified air. **Mechanical ventilations systems … need to be maintained.** **Fresh air supply recommended not to fall below 5–8 litres/second/occupant.** **Aspects of legionellosis also impact here.**
7	Temperature in indoor workplaces	Reasonable temperatures to be maintained. **Should normally be at least 16°C.** **For high temperature situations, insulation, air cooling, shading and siting are all recommended.** Thermometers to be available.
8	Lighting	**Suitable and sufficient lighting to be provided.** **Natural light as far as is reasonably practicable.** **Suitable and sufficient emergency lighting (independent power source) in situations of special exposure to danger in the event of failure of artificial lighting.**
9	Cleanliness and waste material	All workplaces, furniture, finishings and fittings, including surfaces, shall be kept sufficiently clean. Waste materials shall not be allowed to accumulate except in suitable receptacles.
10	Rooms dimensions and space	***Every room where persons work shall have sufficient floor area, height and unoccupied space for purpose of health, safety and welfare.*** **Total volume when empty divided by the number of people normally working in it should be at least 11 cubic metres (minimum). Maximum height dimension to be taken as 3 metres.**
11	Workstations and seating	Workstations to be planned so that each task can be carried our safely and comfortably for all workers, including those with disabilities.
12	Conditions of floors and traffic routes	**Floor and traffic routes should be of sound construction and should have adequate strength and stability, taking account of the loads placed upon them and the traffic passing over them. Floors should not be overloaded.** **Impact here includes conditions of surfaces, steepness, non-slippery, minimisation of likelihood of spillages, minimisation of risks from snow and ice, obstruction-free with particular reference to stairs, steps, escalators, moving walkways, emergency routes, doorways or gangways, near corners or junctions. Applies to parked vehicles.** **Effective drainage to be provided.**

No.	Regulation	Comment
13	Falls or falling objects	**High standard of protection required here.** **Measures are other than personal protective equipment, information, instruction, training or supervision.** **Secure fencing required to exposed edges in respect of falls from height.** **Fixed ladders not to be provided in circumstances where it would be practical to install a staircase.** **Where regular access is needed to roofs, suitable permanent access should be provided.** **Occasional access suggests other safeguards should be provided, e.g. crawling boards, temporary access equipment, etc.** **Changes of level (not obvious) should be marked.** **Provision to be made for work involving loading or unloading vehicles.**
14	Windows, and transparent or translucent doors, gates and walls	Safety material to be used in windows, transparent or translucent surfaces.
15	Windows, skylights and ventilators	**It should be possible to reach and operate the control of openable windows, skylights and ventilators in a safe manner.**
16	Ability to clean windows etc safely	**Suitable provision should be made so that windows and skylights can be cleaned safely if they cannot be cleaned from the ground or other suitable surface.** **Provision includes anchorage points, access equipment, conditions for adequate access, windows which pivot internally.**
17	Organisation, etc. of traffic routes	**Sufficient traffic routes, of sufficient width and headroom to allow people on foot or in vehicles to circulate safely without difficulty.** **Special consideration to be given to people in wheelchairs.** **Suitable measures include, separation of people and vehicles, appropriate crossing points, loading bays, signs, etc.**
18	Doors and gates	**Doors and gates which swing in both directions should have a transparent panel, except if they are low enough to see over.** **Safety features also to be incorporated for sliding doors, upward-opening doors, power-operated doors.**
19	Escalators and moving walkways	**Must function safely, be equipped with necessary safety devices and be fitted with one or more emergency stop controls, identifiable and accessible.**
20	Sanitary conveniences	**Suitable and sufficient sanitary conveniences shall be provided at readily accessible places.**
21	Washing facilities	**Sufficient washing facilities should be provided to enable everyone at work to use them without delay.** **Special provision for workers with disabilities.** **Provision includes protection from the weather, connection to a suitable drainage system, running hot and cold water, privacy, ventilation and cleanliness.**
22	Drinking water	**An adequate supply of wholesome drinking water shall be provided for all persons at work with drinking cups.**
23	Accommodation for clothing	**Suitable and sufficient accommodation shall be provided for clothing not worn during working hours and for special clothing not taken home.**
24	Facilities for changing clothing	**Suitable and sufficient facilities shall be provided for any person at work in the workplace to change special clothing and for reasons of health and propriety cannot be expected to change in another room.**
25	Facilities for rest and to eat meals	**Suitable and sufficient rest and eating facilities shall be provided at readily accessible places.** **Includes seats, rest rooms, rest areas, separate areas for tobacco smokers.**
26	Exemption certificates	**Exemptions may be granted in matters of national security.**
27	Repeals, saving and revocations.	**Extensive list under Schedule 2.**

*For further details and comprehensive coverage refer to: *Workplace Health, Safety and Welfare*, Approved Code of Practice L24, HSE Books.
HSAW Act 1974, Health and Safety at Work, etc. Act 1974.

Appendix Five
DESIGN CHECKLIST

Item	Type of PROCUREMENT (please state …)			
	General	Competence	Design team organigram: are all team members competent or under the supervision of a competent person?	Yes/no
			Can the team competence by suitably demonstrated?	Yes/no
		Resources	Are resources compatible with the nature of the project?	Yes/no
		Co-operation	Have all co-operating parties been identified?	Yes/no
		Co-ordination	Have all co-ordinating parties been identified?	Yes/no
		General principles of prevention	Is the hierarchal approach to health and safety risk management appreciated and endorsed by all team members?	Yes/no
		Procurement	Client's point of contact: Name: tel. no: e-mail:	
			Specialist work packages — 1. — 2.	3.
			Critical dates	
	Procedural	Client	Is the client aware of his duties?	Yes/no
			Documentary evidence as: letter, minutes of meeting, OTHER	Please confirm
		Identification	Have all designers/sub-designers been identified?	Yes/no
			Contact list established?	Yes/no
		Lead designer	Has a '*lead designer*' been identified?	Yes/no
		Design done outside Great Britain.	Is any design being done outside Great Britain?	Yes/no
			If yes, is there an understanding of Regulation 11 duties?	Yes/no
			Understanding confirmed by: interview, letter/e-mail, seminar, training session, workshop.	Please confirm
		Legislation	Does this project have to comply with the Workplace (Health, Safety and Welfare) Regulations 1992?	Yes/no
			Other relevant items of workplace legislation.	Please state
		Design stage	Is there an appreciation of the demarcation between '*initial design*' and '*significant detailed design*'?	Yes/no
		CDM co-ordinator	Appointed?	Yes/no
			Contact details: name: tel.no: e-mail:	
		Contact	Permanent design point of contact: established?	Yes/no
			Temporary design point of contact: established?	Yes/no
			Temporary design point of contact: established?	Yes/no
		Design/construction	Has a design co-ordinator been appointed? Contact details: name: tel.no: e-mail:	Yes/no

	Design	Key dates:	Detailed design:		Tender stage:		
			Drawings:		Pre-construction information:		
			Start on site:		Project handover:		
			Sectional completion		OTHER:		
		Communication	Interfaces with other design elements:				Please indicate:
			Architectural	Fabrication	Heating/ ventilating	Landscape	
			Mechanical / electrical	Structural	Workshop	OTHER:	
		Information	Asbestos management plan	Building log book	Building manual	Condition survey	Please indicate if available or awaited.
			Existing H&S file	Existing drawings	Feasibility	Risk register	
			Services	Structural survey	Traffic	OTHER:	
		Surveys/ investigations	Asbestos	Condition	Contamination	Ground	Please indicate: received or awaited.
			Mining	Party wall	Pollution	Previous usage	
			Radon	Services	Slope stability	Structural integrity	
			Sub-structure	Traffic density	OTHER:		
		Restraints	Building footprint	Delivery routes	Noise	Physical/ adjacent to site	
			Physical/ on-site	Planning conditions	Vibration (ground)	Window of opportunity	
			Working hours	OTHER:			
		Information transfer	Annotated drawings	Brainstorming	Constraints drawing	Design office manual	Please indicate:
			Design philosophy statements	Design report summaries	Design review meetings: minutes	Design risk assessments	
			Design risk register	Notes on drawings	Project risk register	SHE box	
		Loading data	Permanent	Conditions:		Please state:	
			Temporary	Conditions:		Please state:	
		Maintenance methodologies	Developed	Being developed	Strategies awaited	Not relevant	Please indicate:
		Health and safety hazards	Specific hazards identified				Yes/no
			Significant/principal hazards identified				Yes/no
			Hierarchal approach adopted				Yes/no
			Relevant information provided to: client, other designers, contractors, principal contractor, CDM co-ordinator				Please indicate:

Continued

	Stage	Topic					Response
	Construction	Communication	Protocols established between permanent and temporary design				Yes/no
			Temporary works point of contact				Yes/no
			Contact details: Name: tel. no: e-mail:				
		Identification	Have all designers/sub-designers been identified?				Yes/no
			Contact list established?				Yes/no
		Restraints	Access	Boundaries	Compound	Confined spaces	Please indicate
			Ground bearing pressures	Lay-down areas	Loading parameters	Overhead power lines	
			Over-sailing	Sequencing	Structural integrity	Temporary works	
			Temporary stability	Underground structures	OTHER:		
		Design done outside Great Britain	Is any design being done outside Great Britain?				Yes/no
			If yes, is there an understanding of Regulation 11 duties?				Yes/no
			Understanding confirmed by: letter/e-mail, seminar, training session, workshop.				Please confirm
		Information	Has relevant information for the health and safety file been passed to the CDM co-ordinator?				Yes/no
			Certificates	Component weights	DDA	Demolition/ dismantling	Please indicate
			Downloading method statements	Drawings	Eccentric centres of gravity	Energised systems	
			Material hazards	Relevant loading paths	Residual hazards	Structural redundancies	
			Workplace (Health, Safety & Welfare) Regulations 1992 information			OTHER:	
	Post-construction	Inspections/tests	Inspection regimes	access fixings	alarms	control barriers	Please indicate
				electrical systems	fixings (general)	mechanical systems	
				lifts/elevators	lightning protection	safety certificate	
				security systems	structural supports	OTHER:	
			Testing regimes	access fixings	circuitry	control barriers	
				electrical systems	fixings (general)	mechanical systems	
				lifts/elevators	lightning protection	security systems	
				structural supports	OTHER:		
			Major component replacement periods	Component: Period:	Component: Period:	Component: Period:	Please state
			Protective system replacement frequency	System: Period:	System: Period:	System: Period:	
		OTHER:	Please list additional information that should be highlighted in the health and safety file.				

Appendix Six

RIBA OUTLINE PLAN OF WORK 2007 (NOVEMBER 2008 REVISION)

Amended November 2008

RIBA Outline Plan of Work 2007

The Outline Plan of Work organises the process of managing, and designing building projects and administering building contracts into a number of key Work Stages. The sequence or content of Work Stages may vary or they may overlap to suit the procurement method (see pages 2 and 3).

RIBA Work Stages				Description of key tasks	OGC Gateways
Preparation	A	Appraisal		Identification of client's needs and objectives, business case and possible constraints on development. Preparation of feasibility studies and assessment of options to enable the client to decide whether to proceed.	1 Business justification
	B	Design Brief		Development of initial statement of requirements into the Design Brief by or on behalf of the client confirming key requirements and constraints. Identification of procurement method, procedures, organisational structure and range of consultants and others to be engaged for the project.	2 Procurement strategy
Design	C	Concept		Implementation of Design Brief and preparation of additional data. Preparation of Concept Design including outline proposals for structural and building services systems, outline specifications and preliminary cost plan. Review of procurement route.	3A Design Brief and Concept Approval
	D	Design Development		Development of concept design to include structural and building services systems, updated outline specifications and cost plan. Completion of Project Brief. *Application for detailed planning permission.*	
	E	Technical Design		Preparation of technical design(s) and specifications, sufficient to co-ordinate components and elements of the project and *information for statutory standards and construction safety.*	3B Detailed Design Approval
Pre-Construction	F	Production Information	F1	Preparation of production information in sufficient detail to enable a tender or tenders to be obtained. *Application for statutory approvals.*	
			F2	*Preparation of further information for construction required under the building contract.*	
	G	Tender Documentation		*Preparation and/or collation of tender documentation in sufficient detail to enable a tender or tenders to be obtained for the project.*	
	H	Tender Action		*Identification and evaluation of potential contractors and/or specialists for the project.* *Obtaining and appraising tenders; submission of recommendations to the client.*	3C Investment decision
Construction	J	Mobilisation		Letting the building contract, appointing the contractor. Issuing of information to the contractor. Arranging site hand over to the contractor.	
	K	Construction to Practical Completion		Administration of the building contract to Practical Completion. Provision to the contractor of further Information as and when reasonably required. Review of information provided by contractors and specialists.	4 Readiness for Service
Use	L	Post Practical Completion	L1	Administration of the building contract after Practical Completion and making final inspections.	
			L2	Assisting building user during initial occupation period.	
			L3	Review of project performance in use.	5 Benefits evaluation

The activities in *italics* may be moved to suit project requirements, ie:

- D *Application for detailed planning approval;*
- E *Statutory standards and construction safety;*
- F1 *Application for statutory approvals;* and
- F2 *Further information for construction.*
- G+H *Invitation and appraisal of tenders*

Amended November 2008

RIBA Outline Plan of Work 2007

Work Stage Sequences by Procurement Method

The diagrams illustrate different sequences for completion of work stages for various procurement methods, but are not representative of time.

In arriving at an acceptable timescale the choice of procurement method may be as relevant as other more obvious factors such as the amount of work to be done, the client's tendering requirements, risks associated with third party approvals or funding etc.

✣ This symbol indicates that prior to commencement time should be allowed for appointing consultants.

Fully designed project single stage tender

Select advisors	✣ A B Planning
Select / confirm consultants	✣ C D E F1 F2 L3
	G H J K L1+2

Fully designed project with design by contractor or specialist

Select advisors	✣ A B Planning
Select / confirm consultants	✣ C D E F1 F2 F2
Pre-contract design by Specialist	G* H* F1 H2
Post-contract design by Contractor or Specialist	F2 L3
Competitive single stage tender	G H* J K L1+2
Two stage main contract tender	G* H* F1 G H2

G* First stage documentation, H* First stage tender may include Contractor's Proposals, H2 Second stage tender

Design and build project single stage tender

Select advisors	✣ A B Planning Design review
Select / confirm consultants	✣ C D L3
Employer's requirements	G H J K L1+2
Contractor's proposals	E F

Note: final design activity by Client may be at stage C, D, E or possibly F. These stages not repeated by contractor

Design and build project two stage tender (all design by contractor)

Appoint consultants	✣ A B Design review L3
Employer's requirements	G H1 H2 J/K L1+2
Contractor's proposals	C D/E F
	Planning

Partnering contract

	Output specification by client
Appoint consultants	✣ A B Planning
Appoint partnering team	H C D E F1 F2
Select specialists	H E F1 F2 L3
Agree guaranteed maximum price	G/H J/K L1+2

Design and construction sequences may be as shown for Management contract/ Construction management

Amended November 2008

RIBA Outline Plan of Work 2007

Work Stage Sequences by Procurement Method

Management Contract / Construction Management

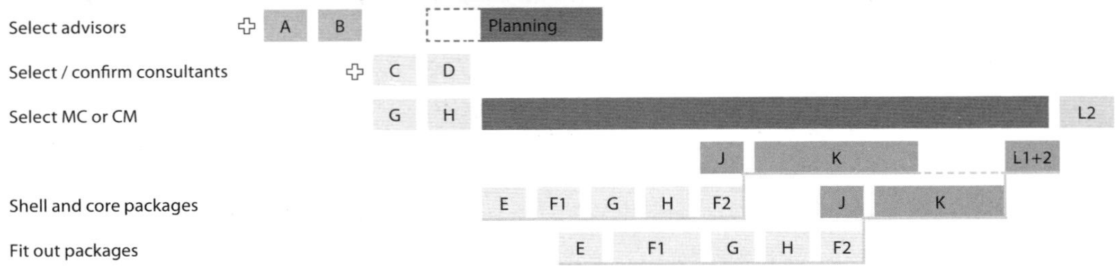

Select advisors — A B Planning

Select / confirm consultants — C D

Select MC or CM — G H L2

J K L1+2

Shell and core packages — E F1 G H F2 J K

Fit out packages — E F1 G H F2

MC = management contractor CM = construction manager

Specialist contractors should be appointed by the management contractor or the construction manager as appropriate in time for the delivery of any pre-construction design services as required by the overall programme. Each package will require building control approval before its construction commences.

Public Private Partnerships and Private Finance Initiative

PPP/PFI stages

Preparation	Tenders/Negotiations	Construction	Use
1.1 Inception	2.1 First bids	3.1 Contract award	4 After hand-over
1.2 Pre-qualification	2.2 Second bids	3.2 Construction	5 Commissioning / operations
1.3 Output specification	2.3 Preferred bidder to financial close		6 Evaluation

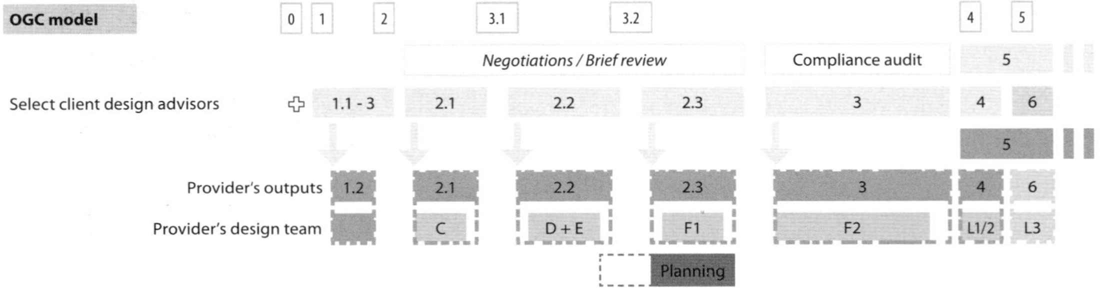

OGC model — 0 1 2 3.1 3.2 4 5

Negotiations / Brief review — Compliance audit — 5

Select client design advisors — 1.1 - 3, 2.1, 2.2, 2.3, 3, 4, 6, 5

Provider's outputs — 1.2, 2.1, 2.2, 2.3, 3, 4, 6

Provider's design team — C, D + E, F1, F2, L1/2, L3

Planning

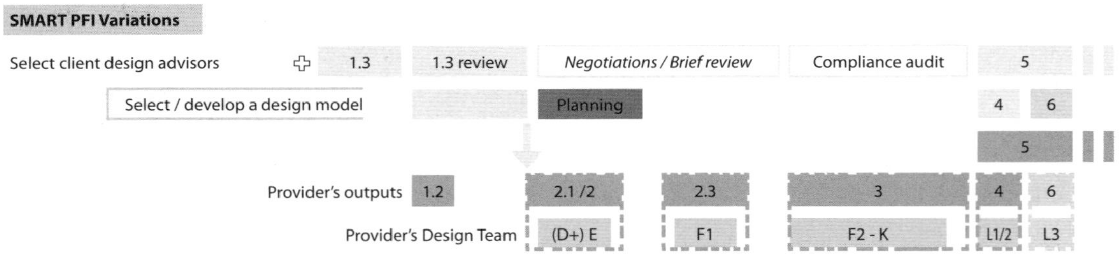

SMART PFI Variations

Select client design advisors — 1.3, 1.3 review, *Negotiations / Brief review*, Compliance audit, 5

Select / develop a design model — Planning — 4, 6, 5

Provider's outputs — 1.2, 2.1 /2, 2.3, 3, 4, 6

Provider's Design Team — (D+) E, F1, F2 - K, L1/2, L3

RIBA Outline Plan of Work 2007 (November 2008 revision) © RIBA (2007). This document is available for free for download from the RIBA's website. Reproduced by permission of the Royal Institute of British Architects, www.architecture.com.

Index

Abbeystead explosion, 8, 9
accountability, 125
ACoP, 37, 38, 39, 40, 42, 63, 64, 70, 72, 73, 77, 81, 87, 88, 116, 125, 126
actions, 28, 31–34, 111
Approved Code of Practice, 18, 38
APS, 58
architect's instructions, 81, 83
archive information, 52, 54
awareness, 67, 78, 80, 132

boundary conditions, 77
Bragg Report, 6, 12–16
brainstorming, 88, 92
buildability, 77, 116
business case, 29, 34

causation, 26
CDM co-ordinator, 25, 27, 32, 37, 41, 47, 50, 51, 56, 72, 73, 77, 82, 83, 105, 111, 116, 124, 125, 126, 129, 132, 133, 134, 135, 139
checklist, 79, 155–158
Cleddau Bridge, 20
client, 25, 26, 29, 32, 47, 51, 52, 54, 56, 59, 63, 64, 66, 70, 72, 73, 77, 81–83, 96, 97, 105, 110, 111, 114, 116, 121, 124–126, 129, 132–135, 139, 142
client's brief, 87
communicate, 78, 80, 81, 104, 105, 121
communication links, 121, 123–125
communication, 52, 54, 67, 70, 78, 81, 83, 87, 88, 105, 116, 121, 123–125, 133, 135, 139
competence, 52, 58, 66, 81, 87, 88, 132
component replacement teams, 47
conceptual, 50
conceptual stage, 77
Confined Spaces Regulations 1997, 37, 42
consequence analysis, 92
consider, 80, 87, 88, 94, 96, 97, 134
consideration, 52, 54, 70, 77, 78, 79, 80, 115, 116, 138, 139
constructability, 47, 63, 77, 78, 79
construction phase, 82, 124, 126, 128, 134, 135
construction phase plan, 50, 51, 83, 104, 105, 125, 133, 142
construction programme, 134
contractor, 9, 16, 47, 50, 51, 52, 56, 59, 63, 64, 70, 73, 77–79, 81, 82, 87–89, 102, 105, 110, 111, 115, 116, 124–126, 128, 129, 132–134, 139
contribution, 52, 54, 67, 78, 79, 80, 81, 97, 105, 111, 129, 135
contributory factors, 19, 20
Control of Noise at Work Regulations 2005, 37, 43
Control of Substances Hazardous to Health Regulations 2005, 37, 44
Control of Vibration at Work Regulations 2005, 37, 43
co-operate, 73, 80
co-operation, 58, 59, 66, 73 , 81, 82, 83, 124, 132, 133
co-ordination, 51, 52, 54, 59, 66, 73, 81, 83, 88, 116, 124, 132, 133
costs, 6, 11, 28, 34

decision trees, 92
definition of design, 63
Delphi techniques, 92
demolition, 134, 138
demolition/decommissioning, 52, 72

design change, 83, 124, 134
design co-ordinator, 139
design duties, 52, 59, 63
design failure, 3, 5–15, 7, 10, 12, 20
design failures, 6
design information, 135, 137–139
design reports, 88, 117
design review meetings, 88, 116
design risk assessment, 88, 89, 101–110, 121
design risk management, 52, 54, 75–84, 88, 89, 105, 121
design statements, 115
design team, 37, 44, 79, 80, 81, 125, 129, 133–135, 139
designer, 25, 27, 32–34, 37, 40–44, 47, 50–52, 56, 58, 63, 64–67, 70–73, 77–79, 87, 88, 96, 97, 102, 105, 114, 124–126, 128, 129, 130–135, 139
desk-top study, 126, 134
documentary evidence, 78, 87–89, 105
documentation, 83, 85–117, 121, 134
duty of care, 40

early contractor involvement, 47
elimination, 70, 79, 80
employer, 28, 73, 89
enabling act, 37, 39
evaluation, 79, 80, 105
examples, 93–98, 101–117

facility managers, 47
failure modes and effects analysis, 92
failure, 3, 18–21, 29, 67, 80, 81, 92, 121
fault tree analysis, 92
forensic analysis, 7, 21
foreseeable, 63, 70, 77, 99, 105
fragmentation, 26, 67

guidance, 37, 38, 41, 58, 59, 63, 78, 89, 129

HASW Act 1974, *see* Health and Safety at Work, etc. Act 1974
hazard analysis (HAZAN), 92
hazard elimination and risk reduction (HERR), 79
hazard identification, 50
hazard and operability studies (HAZOP), 80, 92
hazards, 38, 52, 63, 70, 72, 78, 79, 80, 93–114, 126
hazards in communication (HAZCON), 92
Health and Safety at Work, etc. Act 1974, 33, 37, 39, 40, 43
health and safety file, 50, 51, 52, 54, 72, 88, 105, 111, 115 , 125, 126, 128, 133–135, 137–139, 142
health and safety management, 25, 26, 27, 34, 38, 47, 51, 58, 59, 63, 67, 70, 80, 81, 111, 121, 129
Heathrow Express tunnelling project, 11, 20
holistic, 47, 49, 50, 53, 70, 88, 111

identification, 79, 80, 105, 134
incompetence, 67
information, 66, 67, 70, 72, 73, 79–83, 87, 96, 97, 102, 111, 114, 116, 119–142
initial design, 47, 50, 72, 126, 129, 142
integrate, 38
integrated, 25, 26, 29, 67, 87
  team, 25, 26, 33, 67, 77, 79, 87, 111
integration, 16, 121, 139, 141, 142
interfaces, 58, 81, 83, 129, 132
interventionist, 80, 87

lead designer, 37, 39, 51, 52, 59, 81, 88, 111, 116, 124, 126, 129, 133, 135
likelihood, 80, 104, 126

maintainability, 63, 77, 116
maintenance, 51, 70, 72, 87, 96, 98, 99, 104, 114, 124, 133, 134, 138
maintenance personnel, 47
Management of Health and Safety at Work Regulations 1999, 37, 40, 73, 89
management oversight and risk tree analysis (MORT), 92
Manual Handling Operations Regulations 1992, 37

matrix methods, 92
methodology, 78, 104, 114, 132, 138
MHOR 1992, *see* Manual Handling Operations Regulations 1992

Nicholls Highway tunnel collapse, 12
nomograms, 92
non-notifiable project, 50, 64, 129, 134
notifiable project, 47, 50, 51, 56, 59, 64, 71–73, 82, 126, 129, 133, 135

obsolescence, 81
occupational ill-health, 26, 87
optimum solution, 78
ownership, 67, 83

paperwork, 87, 88, 125
permanent design, 83, 133, 134
planning, 25, 29, 32, 33, 38, 39, 59, 83, 110
planning supervisor, 41
planning/programming, 141, 142
Port of Ramsgate disaster, 6, 10, 20
pre-construction information, 44, 47, 50–52, 66, 82, 104, 105, 115, 124–126, 129, 133, 134, 139
principal contractor, 25, 27, 32, 47, 50, 51, 56, 64, 82, 83, 104, 111, 114, 124, 132–134, 139, 142
principles of prevention, 59, 63, 66, 67, 73, 105
proactive, 25, 29, 33, 58, 77, 80, 81, 111, 125
probability analysis, 92
professional advisor, 34, 37
project management, 25–27, 29, 33, 34, 67, 70, 77, 81, 111, 115, 116, 117, 121
project manager, 27–29, 34, 64, 111
project risk register, 50, 88, 94, 111, 113–114, 121
promptly, 82, 83, 116, 121, 124, 125
proportional, 78
proportionate, 126
public sector, 26

quantum of risk, 70, 77

reasonably practicable, 66, 67, 70, 77, 78, 82
reduction, 79, 80, 87, 110
replaceability, 3
research report, 12, 17, 89
residual, 110
residual hazards, 52, 63, 78, 102, 128, 129, 135, 138
resources, 81
reverse burden of proof, 39
RIBA, 58, 159–162
RIBA plan of work, 159–162
risk assessment, 73, 89, 91–93, 105, 121
risk management, 23, 25, 26, 32, 33, 34, 89, 105, 121
roadmap, 143–144
Robens Report, 37, 39

safe and suitable, 3, 83, 133
sensitivity analysis, 92
severity, 80, 93, 104
significant, 52, 59, 70, 72, 78, 80, 87, 104, 105, 111, 115, 116, 125, 126, 129
significant detailed design, 37, 73, 142
specialist work package, 64, 126, 128
specific, 64, 73, 87, 102, 125, 126
specification, 63, 70, 98–104, 114, 115
stages, 28, 47, 51, 81, 105, 115, 116, 117
stakeholder, 27, 116
statutory duties, 47, 54, 67, 77, 121
sufficiency, 51
sufficient information, 124, 133
Summerland disaster, 6, 8, 20
supply chain, 59, 63, 70, 87, 89
synergy, 29, 33
systems approach, 52–54

tabulation, 88, 89, 92, 93, 111, 121
task analysis, 92
Tay Bridge disaster, 7, 20
temporary design, 73, 133
temporary works, 11, 12
temporary works co-ordinator, 16, 99

temporary works design, 81
temporary works designer, 12, 17, 64
time, 77, 78, 81, 111, 117, 129, 133, 135, 139
time horizon, 52
time management, 26
transfer, 79, 80, 87, 110, 115

usability, 72, 116

variation orders, 51, 81, 83

W (H,S&W) Regulations 1992, *see* Workplace (Health, Safety and Welfare) Regulations 1992
Work at Height Regulations 2005, 37, 42
work-package designers, 47
workplace, 82
Workplace (Health, Safety and Welfare) Regulations 1992, 37, 40, 66, 73, 138, 151–154
workplace risk assessment, 95